THE STANDARDIZED WORK
Field Guide

THE STANDARDIZED WORK

Field Guide

Timothy D. Martin • Jeffrey T. Bell • Scott A. Martin

CRC Press
Taylor & Francis Group
Boca Raton London New York

CRC Press is an imprint of the
Taylor & Francis Group, an **informa** business

A PRODUCTIVITY PRESS BOOK

CRC Press
Taylor & Francis Group
6000 Broken Sound Parkway NW, Suite 300
Boca Raton, FL 33487-2742

Printed in Canada on acid-free paper
Version Date: 20160113

International Standard Book Number-13: 978-1-4987-5201-5 (Paperback)

Library of Congress Cataloging-in-Publication Data

Names: Martin, Timothy D., author. | Bell, Jeffrey T., author. | Martin, Scott A., author.
Title: The standardized work field guide / Timothy D. Martin, Jeffrey T. Bell, and Scott A. Martin.
Description: Boca Raton, FL : CRC Press, 2017. | Includes bibliographical references and index.
Identifiers: LCCN 2015045753 | ISBN 9781498752015
Subjects: LCSH: Workflow. | Standardization. | Production standards.
Classification: LCC HD62.17 .M373 2017 | DDC 658.5/30218--dc23
LC record available at http://lccn.loc.gov/2015045753

Visit the Taylor & Francis Web site at
http://www.taylorandfrancis.com

and the CRC Press Web site at
http://www.crcpress.com

Contents

Foreword vii

Preface ix

Acknowledgments xiii

1 How to Use This Field Guide 1
    Introduction 1

2 Layout Sketch: Where It All Begins 7
    Introduction 7
    Basic Layout Sketch Rules 9
    Some Problems Are More Complicated 11
    Food for Thought 16

3 Questions for Layout Sketch Review 19
    Introduction 19
    Answers for Layout Sketch Questions 27

4 Standardized Work Chart: Building on an Idea 33
    Introduction 33
    SWC 34
    TT and Desired Cycle Time 36
    Stopwatches 37
    Some Tips on Choosing Start/Stop Points 38
    Simple Stopwatch Method 40
    Memory Stopwatch Method #1 42
    Variation 46
    Memory Stopwatch Method #2 48
    Summary of the Three Stopwatch Methods 53
    Food for Thought 55
    Some Other Information on the SWC 56

5 Questions for Standardized Work Chart Review 59
    Introduction 59
    Answers for SWC Questions 68

6 Work Combination Table: Where Time and (Work) Space Collide... 79

Introduction..79

How-To: WCT..82

Meanwhile, Here in the Work (Real) World..91

Real-World Problem: OCT > TT 91

Real-World Problem: OCT < TT (and Forced Wait at End of Work Cycle) 93

Real-World Problem: Forced Wait during Work Cycle 93

Real-World Problem: Showing Work Elements with No Walk between Them..94

A Dose of Reality: Some Problems Are Much More Serious.........................95

Real-World Problem: Work Sequence Does Not Match Geographic Sequence 96

Real-World Problem: Worker Returns to Same Location during Cycle.......96

Real-World Problem: Parallel Machines Due to Excessive Machine Cycle Times #1 98

Some Problems Seem Impossible to Resolve...98

Real-Life Example: Parallel Machines Due to Excessive Machine Cycle Times #2 99

Do Not Get Hung Up: Some Rules Are More Like Guidelines 102

Sometimes, We Outsmart Ourselves 103

Some Food for Thought...105

Watch Your Step: Waste Is Everywhere 106

Real-Life Example: Changing the Unit of Flow during the Worker Cycle......107

Wrap-Up 109

7 Questions for Work Combination Table Review 113

Introduction..113

Answers for WCT Review 120

8 Where Do We Go from Here?...131

Introduction..131

Man–Machine Utilization Graph... 134

Task Summary Sheet..137

9 Questions for Miscellaneous Tools Review....................................145

Introduction..145

Answers for Miscellaneous Tools Questions...151

Reference 157

Index 159

About the Authors 165

Foreword

The real power of Lean occurs when a specialized technique that is typically practiced by a highly technical person such as an industrial engineer is taught and understood and used by an everyday person. I love what Tim and Jeff have created! They have brought standardized work into the realm of understanding of anyone. I do not think that I have seen anyone take a dry subject like this and make it so fun, engaging, and interactive. It kept my inner five-year-old child entertained and wanting to turn each page. It is clearly written and explained in language that mere mortals can understand. The drawings help make the material interesting and fun. The fun and interactive nature of the book helped to embed the learning more deeply. The exercises and the examples are practical and represent real-life situations. I learned a number of tips about standard work that I had not understood before. This is now my go-to book on standardized work.

Joseph E. Swartz
Administrative director, business transformation
Franciscan Alliance, Inc.

In my travels, the concept of standardized work is a topic that people think they fully understand and often downplay its importance or application in the workplace. The variety of scenarios where work is done is broad and highlights the need to document the work so that it can be done consistently by all workers. As a result, people struggle with documenting the best way to do a process and do not implement standardized work as they should. This results in inconsistent output for organizations impacting quality and the effectiveness of their processes.

With this book, I am reminded of the quote that is credited to Taiichi Ohno: "Without standards there can be no *kaizen*." Standardized work is intended to document and improve the work as it applies in all settings, from production to service industries including high- and low-volume/high-mix applications. It is the creation of standardized work that allows for the work to happen with less variation and consistent output.

I first met Tim Martin and Jeff Bell about 10 years ago through a Lean collaborative in our region: the Wabash Valley Lean Network. At the time, they were both working in the automotive industry at Delphi Corporation, previously a subsidiary of General Motors. In their roles, they were part of the continuous improvement group working on customer improvement programs, equipment design, and Lean flow cells.

Since that time, I have gotten to know them well and learned a tremendous amount that has shaped my thinking in my personal Lean journey. I have been able to work alongside them on a Lean implementation at a local hospital and have valued the opportunity to see their passion for improvement firsthand.

Through our time together, I came to learn about their experience, thinking, and improvements utilizing standardized work in a variety of settings. I have heard many stories of the projects and learning that they had with their Japanese sensei, Oba-san. They often recount how many of their lessons were done the hard way, through trial and observing how they did not work, while continuing to learn through the process.

One conversation with Tim about the application of standardized work has stuck with me through the years. It seems elementary, but as we were developing standardized work in a department, we discussed creation of the information for highly repetitive work, which is relatively easy to develop. Also, when tasks are done very infrequently, and it is vital that they are done correctly, standardized work is every bit as important and perhaps even more difficult to develop. The example he shared is the job of changing the toner cartridge in a copier or printer. At best, that task is done every few months, but when it is done, it is very important that it is executed correctly and consistently for it to be effective. The idea struck me, and I have shared that example many times with people in low-volume/high-product mix settings with the message, "We don't do it often, but when we do it, this is the way it is to be done."

This guide is intended to be a workbook that walks you through the development of your standardized work. I encourage you to take it to the *gemba* and use its templates and tables to document your processes as they exist today. It is through its lessons, exercises, and repetitive use that you will gain experience and confidence to develop your documentation and reduce variation to create a better product.

The time Jeff and Tim had with their sensei was invaluable to shaping their ideas and experience with standardized work, for which you are the beneficiary. I know that their mission is to share the lessons and examples with you through their teaching for you to learn and be more effective in improving your products and processes. I hope that you enjoy the book and the lessons as much as I did.

Keep up the good work.

Brian W. Hudson
Senior advisor, Lean Six Sigma
Purdue Healthcare Advisors

Preface

"MEET YOUR AUTHORS AND ADVENTURING PARTNERS"

I never expected to write another book on standardized work after Jeff and I wrote *New Horizons in Standardized Work* (2011). The first book was intended to try to share what we had learned in our many years in the manufacturing industry, attempting to apply standardized work principles, tools, and techniques to a very wide variety of situations, many of which were very different from each other. We had found that these principles, tools, and techniques were not limited simply to the type of processes that they were developed around. On the contrary, the ideas seemed to apply much more widely than simple manufacturing and assembly processes. It was these last points that gave us the courage and drive to write the first book, even though we did not think that anyone would actually read it.

This book is meant to be a field guide—something that readers could take right on out to the floor (or the *gemba*, if you prefer) and refer to it as needed. It is not intended to be a reference guide or anything like that. There are plenty of authors providing some really awesome works on the subject of Lean and standardized work in general. What this book is meant to do is to help readers work through and get used to the methods, tools, and concepts so that they can not only use them but also adapt them to other applications as needed. We also wanted to try and make this journey through standardized work a little more lighthearted.

In order to do this, Jeff and I invited Scott to help us by using his artistic skills and talents to add some humor in the illustrations throughout this book. As you read through this field guide, you will notice that Jeff, Scott, and I have quite a few adventures of our own. This is a separate story in itself, but suffice it to say that it turned out to be a lot of fun for us. We spent many hours working together on the book, but the additional hours we spent on the cartoon illustrations will always stand out in my mind. There were times that we would get to laughing so hard that it was difficult to get back on a serious track. There are also many little added *extras* in some of the illustrations that we will look back on in the years to come and remember the fun times working on this book. This field guide will always have a special place in my heart!

Timothy D. Martin

I will start by saying *thank you* to both Tim and Jeff for allowing me the opportunity of being a part of this incredible adventure. I can honestly say that this has been a life-altering experience for me as both an artist and a student in the realm of standardized work. Throughout the past year of working on this project together, we have traveled through space and time, fought dragons, tamed wild beasts, outwitted wizards, escaped from ninjas, survived countless disasters (both natural and man-made), discovered priceless fortunes, and lived to tell the tales…

In an ironic and evolutionary manner, some of the Lean concepts outlined in this book were actually applied to help create and organize the artwork that you will see as you read. Learning how to *trim the fat* (of unnecessary efforts) so to speak, and streamline both our framework and logistical processes, allowed us to remove the unnecessary steps and focus on delivering the story in a way that best suited the needs of the book, our publisher, and most importantly our own (very unique) sense of humor in an efficient manner. This book has taught me not only how to improve my own creative production output but also how to apply the Lean concept to the other areas of my own personal and professional life.

This has been the single most enjoyable and satisfying work that I have ever done in my life. I only hope that you, the reader, get even a portion of the enjoyment out of the illustrations within these pages as we had in creating them for you.

For their unlimited support and inspiration (and many of the ideas and hidden jokes in the artwork), I would like to thank my family: to my son Mckain, who is my best friend; to my daughter Sara, truly the most talented artist in the family; to my daughter Izzy, who is a never-fading ray of sunshine in my life; to my stepdaughter Katelyn, for accepting me as a part of her life; to my beautiful wife Mahala, who has given me the best 15 years of my life, for her patience and understanding and for believing in me; to my mother Cindi, who has supported my pursuit of art ever since I held my first crayon; to my grandmother *Big Mom*, who always has been and always will be my biggest fan; to my grandfather *Big Dad*, who taught me the *right way* to draw a tree (and I am still trying to get it right); to my sister Skyla, who is the most unique person whom I have ever known; to my aunt Carla, thanks for temporarily adopting me every summer (and allowing me to throw the occasional pool party); to my cousins Amber and Kimberly, who were there for the best days of my childhood; and to my uncle Jason, who first inspired my love of adventure. And, to every one of my other fans, supporters, friends, and extended family, *thank you all for your continued support.* (I originally intended to name you all individually, but I was told that the book needed to be under 300 pages.) Finally, I thank Productivity Press—I hope that this unique approach on teaching standardized work comes as a pleasant surprise.

Scott A. Martin

Hopefully, you will have fun reading this book about standardized work. The authors have done a great job of injecting the lighter side of a topic that is not typically full of fireworks. What better way to help show the lighter side of life than to reflect on history and a little science fiction?

You will see the enforcement of important key points through the extra notes, comments, and follow-up exercises for each chapter. Make sure that you review the illustrations throughout the book for additional insight, which is very difficult to describe in words.

We hope that you gain the insight that you need to help with your journey of establishing a truly engaged workforce in the development of work tasks that will bring you and your business its greatest success time and time again. We also hope that the passion that was used to help bring you this journey into standardized work will rub off on you and your coworkers.

Jeffrey T. Bell

Acknowledgments

We owe a very special "thank you" to several friends and family members whose support, suggestions, and feedback were essential in the completion of this project. Without them, this book would never have been a reality.

George Bell
Marjorie Bell
Cindi Burns
Skyla Burns
Emma Burton
Jason Burton
Kimberly Burton
Katelyn Dodd
Brian Hudson
Amber Jordan
Rebekah Jordan
Todd Jordan
Viktor Jordan
Carla Martin
Jimmy Martin (Big Dad)
Margene Martin (Big Mom)
Mahala Martin
Mckain Martin
Sara Martin
Izzy Martin
Jason Martin
Joe Swartz

In Memory
Dr. James Barany (1930–2011)

Chapter 1

How to Use This Field Guide

Introduction

"THERE ARE MANY WORLDS TO EXPLORE IN LEAN!"

Standardized work is a simple yet powerful concept that can be used in many different areas to help reduce variation, improve quality, and enable more opportunities for improvement. Although it can be applied to both manufacturing and nonmanufacturing processes, it is in the former that the majority of the tools and techniques will be most often used. Therefore, this field guide will primarily feature manufacturing assembly examples as we journey through various aspects surrounding the more common standardized work tools and techniques. Also, in an attempt to cover a very broad range of experiences with standardized work methods, for our readers, we will try to break things down into smaller steps in order to more easily follow the logic that is used to proceed from step to step. Although this may seem a bit tedious to those readers who may already be familiar with many of the principles and methods that are employed, it is our hope that even those with a very deep knowledge of standardized work will find this field guide very helpful.

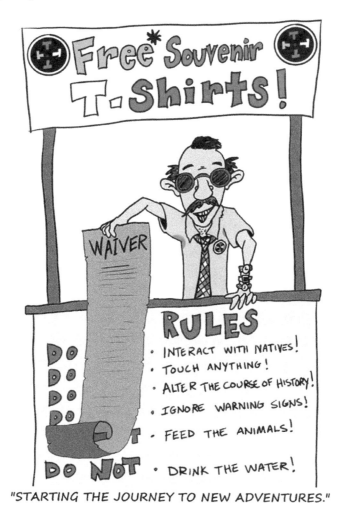

"STARTING THE JOURNEY TO NEW ADVENTURES."

As we get started in our travels, we will explore areas of standardized work that for many are new and unfamiliar. It is our intention to try and make these new areas a little clearer and less difficult to traverse. Over the years, we have noticed a good number of Lean implementations that did not get as in depth into the concept of standardized work as we do in this field guide. We are not saying that they did anything wrong by not delving into many of the areas that are explored in this field guide, but we do feel that these tools and techniques can help us find many more opportunities to improve that might have otherwise been missed.

This field guide is set up to do two main things. First, we will explore several aspects of standardized work in greater depth. While we are at it, we will go through various topics, tools, techniques, and methods that might be of interest to the readers. We will also introduce real-life problems and various other difficulties that they might encounter, mostly from real-life experiences. Although they may not be able to familiarize themselves with all of the types of problems that are discussed during our explorations, they almost certainly will come across some of them during their own adventures in applying standardized work tools and techniques. Hopefully, the problems discussed will offer some extra insight into new problems that you may discover during your own journey.

"WE BEGIN OUR JOURNEY WITH GREAT ANTICIPATION."

Second, we will help the readers become comfortable and familiar with the tools and techniques that are introduced in this book. Simply reading through some examples in the text is not a good way for people to learn, so, after each learning journey, the next chapter is intended to be an opportunity for the readers to work through some problems of their own while using the tools, techniques, and methods in the previous chapter. Hopefully, it will help them to be able to: (a) think deeper about various new problems that might occur, (b) consider what other difficulties might arise if the circumstances were different, and (c) what other issues might offer a challenge to the readers. Working through the problems and issues is not meant to be a chore but rather a learning experience. The problems are not difficult, but they are meant to help the readers to remember the tools and techniques and maybe give them some confidence when trying to apply the tools and techniques to completely unfamiliar applications.

Also, do not worry if you have difficulty solving a particular problem or answering a particular question. The learning chapters are broken into two sections. The first section contains some problems, questions, or other challenges for the readers on the topics that were covered in the previous chapter. The second section contains the answers along with any interim work steps that are required to arrive at the answer.

"NEW ADVENTURES CAN BRING NEW PROBLEMS!"

Along the way in the exercise chapters, we will add additional comments based on the particular question or problem that might offer the readers some insight on how to get started or why the particular problem might be relevant. We will do the same in the answer section since not everyone who reads this field guide will encounter the particular issue or problem, but they might encounter something that is similar. It is our intention that we offer a wide variety of readers from many industries the opportunity to understand and successfully apply standardized work tools, techniques, and general concepts to their own particular industries or business sectors. Therefore, the exercise chapters also have some additional information and insight, so please do not to skip them or else you may miss something interesting.

"WHERE DO WE GO FROM HERE?"

Chapter 2

Layout Sketch: Where It All Begins

Introduction

"GETTING STARTED"

In applications of standardized work where the work is performed in sequential steps at different physical locations as the sequence progresses, the geographic positioning of the various steps required can have a significant impact on the worker. If it is necessary that the worker move on to the next step in the process, then it is imperative that we keep the waste of walking to a minimum. So, the first thing that we realize is that the layout of the process steps is very important, especially in applications where the worker repeats the cycle multiple times per work period.

Let us consider a simple example. In many manual assembly applications, the worker may actually repeat the work cycle hundreds of times per day. If we have, say, 10 feet of unnecessary walking in the job, if the worker repeats the work cycle 100 times per hour (a 36-second cycle) and works in the job for 8 hours, then the amount of unnecessary walk per 8 hours is $100 \times 8 \times 10 = 8000$ feet This is more than 1.5 miles of walk that we have already determined was not needed.

If we assume that the average person takes 0.6 seconds to make a 30-inches step (2.5 feet), then the time taken to walk 8000 feet is 8000 ÷ 2.5 × 0.6 = 1920 seconds or 1920 ÷ 60 = 32 minutes of the 8 hours. If we were somehow able to separate this out from his or her normal work time and did this same job for an entire year (let us say 250 days), the total time used for just this extra 10 feet of walk per cycle would be equal to 32 × 250 = 8000 minutes or 8000 ÷ 60 = 133.3 hours. End to end, that is more than 16 8-hour days spent on just walking those unnecessary 10 feet each cycle.

On top of all this, consider that very often the worker is carrying a part from step to step and that these parts have some weight that is associated with them as well as some odd shapes that might not be designed with ergonomic considerations in mind. Upon deeper reflection, we notice that we are paying the worker for the time to walk 8000 feet, and all he or she is doing is walking while carrying a part. Not only is this unnecessary; the worker is also needlessly fatigued (see Figure 2.1).

We now see that the excessive distance between successive work steps can greatly impact the worker as well as add waste to his or her job. If we are not diligent in our efforts of job design, we can easily add enormous waste into the work. At this point, it becomes evident that the layout plays an important role in standardized work.

Before we can analyze a layout of process steps for improvement opportunities, we first must capture the layout in a meaningful way. Once we have done this, we can also use it for communicating with others. Although for some

Figure 2.1   Avoid unnecessary work whenever possible.

Figure 2.2   Basic layout sketch rules (guidelines actually).

communication purposes, we may need a very accurate representation, in the beginning, we only need a sketch that reflects the basics of the layout. We simply refer to this as a layout sketch. It is very easy to produce a layout sketch. In order to help move things along, we have provided some *rules* (Martin and Bell 2011, p. 38) that we find helpful. (Although we may refer to them as rules, they are not really carved in stone—they are more like guidelines, as shown in Figure 2.2.)

Basic Layout Sketch Rules

- Sketch a rough approximation of each workstation and its location in relation to the other workstations.
- Each major work sequence step is shown as a number within a circle.
- The steps are connected by solid arrows.
- The return to the step that begins to repeat the sequence is shown with a dashed arrow.
- The work step, equipment name, etc., can be shown for clarity.

As we get started in our creation of a layout sketch, we need an example application to use. For most of the examples in this field guide, we will use the assembly of a simple toy truck (Figure 2.3). There are five major steps in this example:

1. *Body assembly*—putting the body parts onto the chassis
2. *Wheel assembly*—adding the wheels to the body assembly
3. *Final assembly*—adding the people and the top to the wheel assembly
4. *Wheel test*—placing the final assembly into a *wheel tester* machine
5. *Inspect and pack*—inspecting the tested assembly for defects and packing if good

Figure 2.3 Truck example.

In our truck example, we will refer to the various locations as workstations or just stations for short. From our discussion on the waste of excess walking in the "Introduction" section, we know that we need to keep the distance between successive steps at a minimum. Let us say that the stations are simply placed side by side in a line and the sequence goes from right to left, as shown in Figure 2.4.

Before we go any further, we see at least one problem jumping out at us. Although the next station in the sequence is very close to the previous one minimizing the walk between them, this does not hold true if the worker continues repeating the same cycle over and over. Once a cycle is complete, he or she must walk back to the first station to start the next one. So, in this layout, although

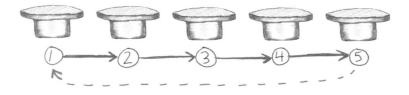

Figure 2.4 A work layout with station side by side in straight line.

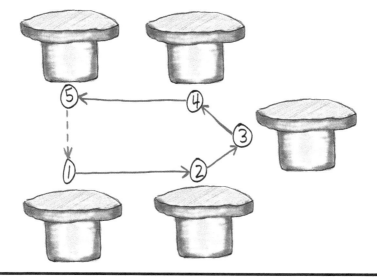

Figure 2.5 Same work as previous diagram, but in U-shaped work cell layout.

the walk between consecutive steps seems minimized, this is not the case when going from the last step back to the first step. In order to reduce this distance, the last station must be closer to the first station. As this distance is reduced, instead of a straight line like what we started with, the layout must be curved or folded. Based on the size and shape of the stations (often referred to as their footprint), the layout will take on a more folded shape—this is why these layouts, most often called work *cells*, are shaped more like the letter *U* (see Figure 2.5).

Some Problems Are More Complicated

Another important point to consider before we go on with our updated example is that the worker moves in a kind of circular pattern in a clockwise direction. If the layout was in the opposite direction, some might say that he or she could also travel in a counterclockwise (or anticlockwise) direction just as easily. Which one is best? To answer that, we must take a few moments to consider things from a broader perspective. First, it is a fairly common assumption that the majority of people in the world are categorized as right handed. What impact will this have on a work cell layout? As we start to think deeper about this, it is also important to consider the size and shape of any items that the worker may be required to carry between stations—and whether or not they require both hands or only one hand. All of this could keep us bogged down for a very long time trying to develop the perfect layout for every work cell that comes along. Fortunately, there are a few guidelines that we can use that will help us.

"A COMPLICATED PROBLEM."

In our experiences, the best way to describe this is to first consider the clockwise-versus-counterclockwise issues. If the worker has both hands in use upon approach to a station (carrying a part, for example), the question seems to be irrelevant as the situation appears to be the same regardless of the direction of approach. In such a situation, he or she would have to stop, in some cases turn, and then place the part using both hands (assuming that the machine is empty), so neither direction of approach would seem to have an advantage over the other at this point. So, for the moment, we will stop our analysis of two-handed carry scenarios to look at one-handed scenarios—does one have an advantage over the other? See Figure 2.6.

Figure 2.6   Is one hand better than the other?

As we try to answer this latest question, we must consider what the worker will be doing when he or she arrives at the station. In some instances, he or she will stop, possibly turn, place the part down, do some additional work, pick the part back up, and then move on to the next station. If this is the case, there may be no significant advantage for this particular station. In other instances, the worker may only remove (unload) a part that was left at the station from the previous cycle, load the new part, and continue—this occurs very frequently in order to eliminate the waste of having him or her wait for something to complete (such as a machine operation). If the hand closest to the station is occupied upon approach, then the situation is similar to that where both hands are occupied, and thus this seems to offer no advantage. However, in this scenario, there appears to be an advantage if the hand nearest the completed part from the previous cycle is unoccupied because this offers the opportunity for the unoccupied hand to start unloading the completed part earlier and thus reduce the time that is spent at the station by the worker.

"BORING, BUT NECESSARY ANALYSIS."

Now that we have surmised that there is an advantage to having the leading hand free upon approach, there are a couple more things to consider. First, which hand requires the most accuracy—the one unloading or the one loading? The answer is easy as this is what defines the right- or left-handed category of the worker. Since loading almost always includes some form of part location (pins, rails, etc.), a bit more dexterity is required in loading than unloading, which usually is just a single simple motion. Thus, the dominant hand (right or left scenario) is the one that most workers will use to load with—which means that the nondominant hand is the one that we would want to be unoccupied and leading upon approach to the station. When this is the case, the unloading can begin as soon as the unoccupied hand is within range so that the part in the occupied hand can begin loading as soon as it is within range. In instances where no other work is required by the worker, he or she may not even have to stop and turn, thus eliminating even more time.

One last point, once the worker leaves the station in our scenario, the dominant hand is unoccupied because the part is still in the nondominant hand. This means that he or she will need to move the part to the other hand before arriving at the next station. However, since this can be done during the walk to the next station, it is not a significant issue. As we complete our analysis, we realize that upon approach, there are only two combinations where the leading hand is unoccupied and the dominant hand is occupied: (1) right hand/counterclockwise and (2) left hand/clockwise. Since the consensus is that the majority of people are right handed, in this field guide, we will go with the assumption that the cell direction will be counterclockwise to accommodate the largest number of workers.

"PUTTING OUR ANALYSIS TO GOOD USE."

Now that the preferred direction of process step progression has been established, we can proceed with our discussion on layout sketches.

DRAW A ROUGH SKETCH OF THE LAYOUT

To create a layout sketch, it is important to understand the geographical perspective of the stations and their relationship with the sequence of the work flow. First, draw a sketch that represents the geographic layout of the work area. Show only the equipment that is a part of the worker's cycle at this point.

"SKETCH THE GEOGRAPHIC LAYOUT OF THE WORK AREA."

SHOW FLOW DIRECTION USING NUMBERS IN CIRCLES AND ARROWS

Next, using a number inside of a circle, label each process step location in the work sequence as it is performed even if the worker must walk back and forth or even retrace steps. Then, use arrows to show movement from one location to another—solid line arrows for steps that are internal to the sequence and a dashed line arrow to show when the operator has completed an entire cycle and is about to start the next complete cycle. If several steps are generally at the same location, consider that as a single geographic location for most instances, and number the steps accordingly—it is possible that a location has more than one sequence number.

"SHOW THE FLOW AND GEOGRAPHIC LAYOUT USING NUMBERS AND ARROWS."

In order to keep the layout sketch as simple as possible, it is important to only include information that is relevant to the work and necessary to show any problems that are caused by outside issues. For example, if there are physical obstructions—things that would cause the worker to vary his or her path in order to get to the next workstation—these should be included as they are necessary to help illustrate a problem that exists.

Food for Thought

In our discussion on cell direction in the "Some Problems Are More Complicated" section, we stopped our analysis of the scenario where the worker was required to use both hands to carry a part to the next station. At that point, the direction of approach seemed irrelevant to our discussion. However, there are some important issues to consider in situations where a two-handed carry is required.

In some instances, the worker will stop, turn, place the part down, do some additional work, pick the part back up, and move on to the next station. In such situations, no part is left in the station, so, upon approach, he or she has somewhere to place the part. But, what about instances where he or she may only remove (unload) a part that was left at the station from the previous cycle, load the new part, and continue? Remember that, upon approach, the worker is already carrying a part. In order to place the new part, the completed part may need to be removed. This presents some additional issues.

"FOOD FOR THOUGHT."

In this scenario, if the worker cannot load the part that he or she is carrying directly into the machine, then he or she must first place the part in an interim spot, unload the previously completed part and place it at yet another interim spot, then pick up the first part, place it in the machine nest, and then pick up the previously completed part before moving on. This scenario can involve an enormous amount of wasted motion if we are not diligent in our efforts at designing the cell.

There are several ways to eliminate or reduce the wasted motion that was described above. Unfortunately, they all seem to add additional cost and complexity. Some that we have seen include automatic unloading of a completed part, dual nesting, and using a second machine. Out of these three options, the one we like best is the automatic unloading. In this scenario, the nest is empty upon approach, so the worker can place the new part, grab the previously completed part, and move on. The dual nesting scenario usually requires a lot more complexity than the automatic unloading and thus can often be very costly. Using a second machine seems to be a very costly solution as well; however, it is seen more frequently than you might expect.

"TIME TO MOVE ON FROM LAYOUT SKETCHES."

Chapter 3

Questions for Layout Sketch Review

Introduction

"PROBLEMS CAN ARISE AT ANY TIME!"

In Chapter 2, we walked through the basic process of how to develop a layout sketch. We also took a few side trips to discuss some of the closely related issues that surround the layout of a work cell. The intent of doing it in the manner we did was to also introduce the idea that continuous improvement is not just something that you do after the fact. If we learn to think deeper about what we are doing as well as why we are doing it, it is possible to begin the improvement process much earlier. If the cell already exists, we can begin to identify areas where we can start improving. But, if the cell does not yet exist or is still being designed, we might possibly start making improvements even before the equipment is acquired or the work is defined.

We have found that the best way to learn is by doing. But, just simply repeating the steps for doing something does not help us to think deeper about it. There must be other aspects that are introduced in order to take us out into a little more unfamiliar territory so that we are forced to stop and think. It is in this way that we are placed in the position of trying to understand the impact that changing the conditions or variables has on the particular situation that we are learning about. The best way is to be right there at the *gemba* (where the real work is occurring). However, if the cell and the work are still being designed, or if the workplace has not yet been assembled, the situation is much different. In this case, the *gemba* is wherever we are doing the designing. Sometimes, this is in an office environment or in a work area where we are trying to fabricate and use full-size mock-ups. No matter the environment, we can still go to the *gemba* and, from there, continue to learn and think deeper about the situation so that we can strive for continuous improvement, even before the problems actually exist.

"STABILIZE THE SITUATION BEFORE STARTING THE IMPROVEMENT PROCESS."

Question 1: A worker has 12.5 feet of walk that is considered excessive in a work cell layout. How many miles per year would this be equivalent to?
Given

1. The work cell has a 36-second cycle time.
2. The work time is 7.5 hours/day.
3. There are 250 workdays per year.

Remember: 1 hour = 3600 seconds
         5280 feet/miles

Use the answer area for question 1 for your answer.

```
┌─────────────────────────────────────────────┐
│                                             │
│                                             │
│                                             │
│                                             │
│                                             │
│                                             │
│                                             │
│                                             │
└─────────────────────────────────────────────┘
```

ANSWER AREA FOR QUESTION 1

When the number of cycles required for a work cell per year is high, even a single foot of unnecessary walk can have a huge impact. If the extra walk is significant in itself, the impact can often be dramatic. Sometimes, it is difficult to visualize the impact that waste has on a situation. When dealing with walking distances, we often found that a useful way to help others understand the impact is to convert the annual total of the extra walk from something that they could compare to something that they could more easily relate to. Sometimes, just using miles of walk per year can make a lasting impression. Other times, it might be to use some other familiar distances such as football fields, buildings, mountains, or the distance from one city to another.

It is important to note that even though we can easily appreciate the impact of making improvements, we often have to get buy-in and approval from others before we can implement the improvement. This is especially true when we need to spend money or resources in order to try the improvement. Therefore, we may have to try and find a way to help someone else more easily compare the cost of implementation to the benefits that are expected from the improvement. Another way that we have found useful is to put the expected benefits into added capacity, revenue, and output instead of reduced costs.

Question 2: If 12.5 feet of extra walk could be eliminated from the work cell layout in question 1, how many more parts per year could be made in the work cell?

Given

1. One step of walk is equal to 2.5 feet
2. The worker can travel 2.5 feet or one step in 0.60 seconds.

Use the answer area for question 2 for your answer.

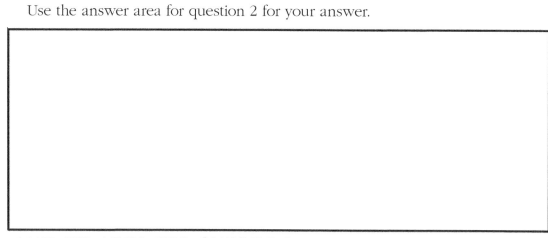

ANSWER AREA FOR QUESTION 2

In the first exercise, we tried to visualize the impact of unnecessary walk, which we know would reduce labor costs. However, in the second exercise, we find that another useful way to look at improvement opportunities is to show the impact on output in terms that are easily understood. Increased capacity is very helpful, especially when we are dealing with situations where the more parts we can produce, the more revenue we can generate (which assumes of course that the parts can be sold).

"SOME PROBLEMS REQUIRE US TO ADAPT TO THE SITUATION."

Finding a way to express the added expenses or the lost opportunities due to waste in a work cell is usually a very important part of getting everyone focused on continuous improvement. It can have an enormous impact on those who are working in the cell as well. Eliminating unnecessary walk not only helps reduce wasted expenses; the worker is also not as tired at the end of the day, and that makes his or her life easier. However, if the worker is kept engaged and involved in making the work better, it can have an even greater impact over time as most workers get incredible satisfaction from helping to improve their own job. Most people take great pride in the work that they do, and getting involved in continuously improving the work can be a great staff satisfier—and, who knows, the work might be better than those who do it every day? It can definitely mean a win–win situation when the workers and the company are working together to make things better.

"WE LEARN BY DOING-PRACTICE IS THE KEY!"

Question 3: (Choose one) Which of the following steps is incorrect when creating a basic layout sketch?

A. Each major work sequence step is shown as a number within a circle.
B. The steps are connected by solid arrows.
C. The return to the step that begins to repeat the sequence is shown with a solid arrow.
D. Work step, equipment name, etc., can be shown for clarity.

Use the answer area for question 3 for your answer.

ANSWER AREA FOR QUESTION 3

Question 4: In a two-handed scenario, what would happen if the machine was not empty when the worker arrived?

Use the answer area for question 4 for your answer.

ANSWER AREA FOR QUESTION 4

These situations are common for work cells where there is a machine cycle that is involved. If the machine cycle is very short, it may be worthwhile to just let the worker wait for the machine to complete the cycle so that he or she can unload the part and move on. However, if the machine cycle is not very short, it is often undesirable to have the worker wait. In order to allow the worker to move on, he or she would need a part that had already been completed. This means that the part from the previous cycle would be complete and ready to take in the place of the one being brought by the worker on this cycle. It also means that there must always be one part of work-in-process inventory for the cycle to be the same each time.

"SOMETIMES BOTH HANDS ARE NEEDED."

It does not take a very in-depth analysis to figure out that if both hands are occupied upon approach, before the worker can do anything else, he or she must free up his or her hands. The next logical step is to set the part(s) down. If he or she is able to set the part directly into the station nest, then his or her hands are free again, and work can continue. However, if the worker is required to place the part at an intermediate location so that he or she can manually move the part in the work nest of the station, he or she also has to do the same with the part that he or she is unloading—put it in an intermediate unload spot so that it can be picked back up upon leaving for the next station in the work cycle. All of these intermediate moves are waste as we well know. The best solution would be to have an automatic unload after the machine is complete so that as the worker sets down the part that he or she is carrying, it can be placed directly into the machine nest so that the cycle can start as soon as possible. The worker can then pick up the part from the previous cycle and move on to the next station.

"THINKING DEEPLY IS OFTEN REQUIRED TO FIND WHAT WE ARE LOOKING FOR."

Question 5: (a) (True or false) When designing the layout for a work cell, establishing a counterclockwise direction of flow in the work cell would be the preferred direction for the typical work force. (b) Also, briefly explain why your answer is either true or false. Use the answer area for questions 5a and 5b for your answer.

ANSWER AREA FOR QUESTIONS 5a AND 5b

Question 6: (Choose one) Which of the following options is the most expensive one for establishing an empty work nest for a workstation or a machine in a work cell?

A. Automatic unload for a completed part
B. Dual nesting feature
C. An additional machine
D. A duplicate work cell

Use the answer area for question 6 for your answer.

```
┌─────────────────────────────────────────────────────────────────┐
│                                                                   │
│                                                                   │
└─────────────────────────────────────────────────────────────────┘
```

ANSWER AREA FOR QUESTION 6

When we start to think deeper about the problem of a work cell that has a machine process time that is too long for the worker to stay and wait for the cycle to complete, our thoughts quickly progress to considering the best alternative for dealing with the issue. As often happens in a work cell, the machine process does not always take into account the problem that we are considering. In such cases, we are faced with finding the best way to resolve the issue in our particular situation.

"THE SEARCH FOR A BETTER WAY IS NEVER-ENDING!"

If the machine manufacturer does not offer a solution, we may be forced to consider other less desirable alternatives. Adding an automatic unload or a dual nesting feature to a purchased machine or designing and building a custom machine will require engaging a custom equipment builder. Not only can this be very expensive; it will also add additional problems due to unforeseen issues of the customizations—deeper thinking by the machine builder is critical, and we cannot always count on this covering every aspect of the process. The cost of an additional machine (or machines) can actually be the lesser of the two evils—the cost of off-the-shelf equipment versus the costs of overly complicated modified or custom-built equipment and added downtime due to modification-related issues.

Answers for Layout Sketch Questions

Question 1: A worker has 12.5 feet of walk that is considered excessive in a work cell layout. How many miles per year would this be equivalent to?

"SOLVE YOUR PROBLEMS ONE BY ONE."

Given

1. The work cell has a 36-second cycle time.
2. The work time is 7.5 hours/day.
3. There are 250 workdays per year.

Remember: 1 hour = 3600 seconds
5280 feet/miles

See answer for question 1.

> a) 3600 seconds per hour divided by 36 seconds cycle time = 100 work cycles per hour
> b) 7.5 hours per work day multiplied by 100 work cycles per hour = 7,500 work cycles per work day
> c) 7,500 work cycles per work day multiplied by 250 work days per year = 187,500 work cycles per year
> d) 187,500 work cycles per year multiplied by 12.5 extra feet per work cycle = 2,343,750 extra feet per year
> e) 2,343,750 extra feet per year divided by 5,280 feet per mile = 443 miles per year (16+ marathons of extra walk per year for the worker!)

ANSWER FOR QUESTION 1

A marathon is about 26 miles and some odd feet. In this exercise, the number of added miles per year that the worker must walk is converted to the number of full marathons that would be represented by the distance. This might have a good impact on a certain audience and less of an impact on others. The important issue is to find a way to relate the distance to your particular audience. Often, people will put the distance in several different contexts in order to try and reach multiple audiences simultaneously. The comparison used is not as important as getting the audience to understand the impact.

"THINK DEEPLY AND TRY YOUR IMPROVEMENT IDEAS."

Question 2: If 12.5 feet of extra walk could be eliminated from the work cell layout in question 1, how many more parts per year could be made in the work cell?

Given

1. One step of walk is equal to 2.5 feet
2. The worker can travel 2.5 feet or one step in 0.60 seconds.

See answer for question 2.

a) 12.5 feet of extra walk per work cycle divided by 2.5 feet per step = 5 steps of extra walk per work cycle

b) 5 steps of extra walk per work cycle multiplied by 0.6 seconds per step = 3 seconds of extra walk time

c) 36 seconds work cell cycle time − 3 seconds of extra walk time = 33 seconds is new work cell cycle time

d) 3,600 seconds per hour divided by 33 seconds work cell cycle time = 109 work cycles per hour

e) 109 work cycles per hour multiplied by 7.5 hours per work day = 817 work cycles per work day

f) 817 work cycles per work day multiplied by 250 work days per year = 204,250 work cycles per year

g) 204,250 new work cycles per year − 187,500 previous work cycles per year = <u>**16,750 work cycles per year increase**</u>

ANSWER FOR QUESTION 2

Putting the lost opportunities into the right context is very important as we have discussed in the "Introduction" section. In this exercise, the added capacity can mean more revenue, less overtime, and so on. Understand that continuous improvement is not simply a way of saving the company money but also a way to preserve jobs, and growing the business can be a very high motivator—especially in these tough economic times. Continuous improvement also gives the workers a chance to try and get involved with helping preserve their own jobs as well as the jobs of their coworkers. Most importantly, it gives everyone the chance to be part of the team.

"ALWAYS LEARN FROM YOUR MISTAKES!"

Question 3: (Choose one) Which of the following steps is incorrect when creating a basic layout sketch?

A. Each major work sequence step is shown as a number within a circle.
B. The steps are connected by solid arrows.

C. The return to the step that begins to repeat the sequence is shown with a solid arrow.

D. Work step, equipment name, etc., can be shown for clarity.

See answer for question 3.

> **The answer is "C".** The return to step that begins to repeat the sequence is shown with a dashed arrow.

ANSWER FOR QUESTION 3

Question 4: In a two-handed scenario, what would happen if the machine was not empty when the worker arrived?

See answer for question 4.

> The worker must have a designated location to place the to-be processed part with both hands safely before getting the finished part from the machine. The worker would also need a designated location to place the finished part before getting the to-be processed part to load to the machine.

ANSWER FOR QUESTION 4

"SOMETIMES WE JUST HAVE TO START OVER."

As we should be able to recognize at this point, the parts that require the worker to use both hands to carry can add extra complexity to the work cell if there are machine cycles involved that require the worker to leave a part in the machine and continue on with a part that was left from the previous cycle. This was not an issue with parts that only required one hand as the worker could use both hands at the same time—to unload with one hand while simultaneously unloading the completed part with the other hand. Understanding this can help make a great impact on knowing the things that drive cost during the design phase. It might not be possible to have that much impact on the size of a part, for example, but for those parts that are awkward to handle, it could make the difference between a one- and a two-handed carry. Sometimes, being creative in the overall shape, center of gravity, footprint, or other features involved with how the part is handled by a person can make an enormous difference.

"PROGRESS CAN SOMETIMES BE PAINFUL."

Question 5: (a) (True or false) When designing the layout for a work cell, establishing a counterclockwise direction of flow in the work cell would be the preferred direction for the typical work force. (b) Also, briefly explain why your answer is either true or false. (See answer for questions 5a and 5b.)

> **The answer is True.** For two-handed loading/unloading of parts, the direction of flow does not usually matter. When one-handed handling of parts can be achieved, a counter-clockwise direction of flow would support a predominately right-handed work force. This would allow for the additional dexterity of safe control of material handling.

ANSWER FOR QUESTIONS 5a AND 5b

Counterclockwise would be the preferred direction for a typical work force or if we wanted to make sure that we design the layout to accommodate the largest number of workers. However, it is conceivable that we would want to design the cell for a single individual. This often happens in smaller businesses where the worker turnover is not expected to be an issue. Another situation where this could be the case is where the workstations can easily and quickly be rearranged. In such a scenario, the station nests and the equipment would have to be designed to be used for either direction.

Question 6: (Choose one) Which of the following options is the most expensive one for establishing an empty work nest for a workstation or a machine in a work cell?

A. Automatic unload for a completed part
B. Dual nesting feature
C. An additional machine
D. A duplicate work cell

See answer for question 6.

> **The answer is D.** A duplicate work cell would not only add twice the overall work cell cost, but would also duplicate the additional worker wastes of walk, wait and floor space.

ANSWER FOR QUESTION 6

"MOVING ON TO THE NEXT STEP!"

Chapter 4

Standardized Work Chart: Building on an Idea

Introduction

"BUILDING ON AN IDEA OF CONTINUOUS IMPROVEMENT."

In Chapter 3, we discussed the importance of the layout sketch. We learned that it is essential to show not only the problems and issues that are associated with the geographic positioning of the workstations but also how the worker will interface with the stations. With proper planning and deeper thinking, it is possible to use a good layout sketch to start making improvements early—in some cases, even before the work cell is built.

So, how do we build on this foundation—the idea of continuously making improvements as we go along? We can do this by continuing to think deeper about what it is that we are trying to accomplish and the steps that we are taking for the objectives. In the case of the layout sketch, we know that it is an important tool, but how do we support the idea of continuously searching for ways to do things even better? For a start, we must ensure that the layout sketch is one of the basic starting points in our efforts to improve. A way that we can do this is by creating a standardized work chart (SWC). But, what exactly is a standardized work chart?

"JUST WHAT IS A STANDARDIZED WORK CHART?"

SWC

As we discussed in *New Horizons in Standardized Work* (Martin and Bell 2011, p. 2), standardized work can simply be described as "the currently best-known method for accomplishing the work." It does not say that it is the only way to do the work. There are usually many different ways to accomplish the work. But, if different workers are doing the work differently, the results will not always be the same, and, quite often, the amount of time taken to do the work is not the same either. When the results are not exactly the same every time, we describe the differences as variation. We have already learned that reducing unnecessary activities such as walking is a fundamental step in making improvements; reducing variation is another. In our discussion in Chapter 2 on cell layout design and analysis, we determined that in order to accommodate the greatest number of workers, most work cells are designed for right-hand workers progressing along a counterclockwise path. In order to communicate the details of how the work in the cell is intended to be performed, as well as serve as a basis for continuous improvement, we must capture and document the information in a clear and consistent fashion. This is one of the purposes of the SWC. A simple SWC is shown in Figure 4.1.

The SWC should be very simple and only contain information that is relevant for communicating the current method that is to be used for accomplishing the work for a single-worker job. The central point of the SWC is the layout sketch. This conveys the sequence and direction of the worker when performing the work. However, it does not tell us how long the work should take if done correctly. And, since each worker is different, how do we determine how long we think that the work should take when performed correctly? For this, we need a standard time for comparison purposes on the SWC as well. But, how do we determine this standard time? In *New Horizons in Standardized Work* (Martin and Bell 2011, pp. 59–64), we discussed the concept of takt time (TT). In the simplest terms, takt time "is used to describe the cadence or pace at which a product (or service) needs to be produced in order to satisfy customer demand in the time we have allotted to do so." This tells us how often something should be completed in order to meet our customer's requirements based on our planned working schedules, not how long the work should actually take. We will need to think deeper if we want to find a way to determine this time.

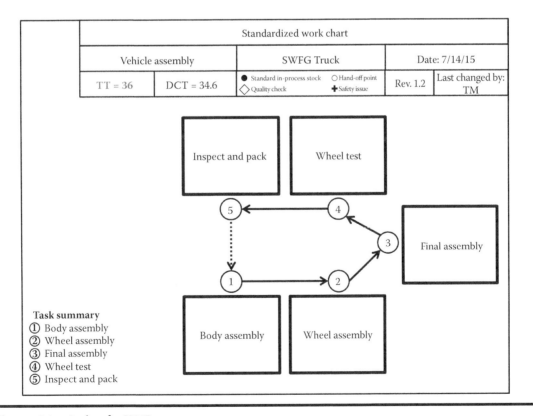

Figure 4.1 A simple SWC.

"HOW LONG SHOULD THE WORK TAKE?"

TT and Desired Cycle Time

If we observe workers doing the work correctly and measure the time that is taken, it is possible to determine a time that we feel is a reasonable target or, in other words, the desired cycle time (DCT). The DCT should be less than or equal to the TT; otherwise, we will not be able to meet our customer's requirements. TT is calculated by dividing the customer requirements for a period of time (day, week, month, etc.) into the work time that is allotted for meeting these requirements during the same time period. For example, if our customer needs 140 parts per day and we have dedicated 420 minutes of work time per day, then

$$TT = 420 \text{ minutes} \div 140 \text{ parts} = 3 \text{ minutes/part}$$

However, this does not tell us that it takes 3 minutes of work to complete a part; it simply says that in order to meet our customer's requirements within the planned work time, we will need to produce at a rate that equals 3 minutes per part—if we intend to meet the requirements with 420 minutes of work time per day.

If the DCT and TT do not match for our planned schedule, then we must adjust the working time that is allotted and/or the number of workers/cells to vary the TT in order to achieve an acceptable match between the time that is taken by the worker(s) and the customer requirements. The DCT is based on establishing a standard time for the work itself regardless of the variations in the schedules. Also, there may be multiple workers who are involved in a work cell, and their individual DCTs may not always be equal—even though the TT will be the same for all of them. For this reason, we prefer to include both the DCT and TT on the SWC along with the layout sketch. We can easily calculate the TT, but how can we determine the DCT? First, we must observe, but we know that even the same worker will have some variation, so we will need a way to choose what we feel is the best time to use for the DCT. Before we can do that, we need to find a way to measure the time that the worker takes for each cycle—for this, we will need a stopwatch.

Stopwatches

There are two main types of stopwatches (see Figure 4.2). The simplest one just displays the time that is accumulated from the initial actuation of the start/stop button to the next successive actuation where the accumulation of time stops and a way to reset the timer to zero. This is good for a single event, but it is not adequate for measuring consecutive events since it stops the timing on the second activation of the button. The other type has a feature that is referred to as memories. This feature allows the user to store the time between successive actuations of the button, while the stopwatch continues to accumulate the time from the initial actuation of the button until the timer is stopped or until the maximum number of memories is reached. This type of stopwatch normally has two or more rows that display time. Usually, one of them will display the continued

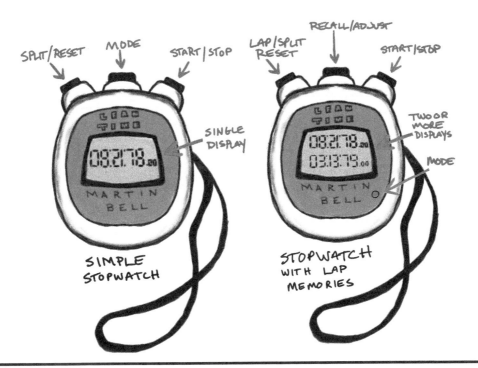

Figure 4.2 Typical examples of the two most common stopwatch types.

accumulation of time, while the other will indicate the time that is captured since the last timing actuation. Some have more than two displays, but they are not critical to our discussion. Regardless of the number of displays present, after the observation is complete, the memories can be recalled in the original order so that it is possible to measure sequential events of greatly varying times—when done, it also has a way to reset the display and memories back to zero. The first type may seem too simple for our purposes yet can still be useful to us. But, before we go on, it is necessary for us to think deeper about what we are trying to measure and how we will do it.

In our discussion about the DCT, we found that we need to determine a target (or standard) time that the work should basically take each and every time that it is performed correctly. But, if there is some variation, even between cycles by the same worker (we are only human after all), how do we determine the best number to use as the standard time? There are several methods that we can use, such as average or lowest repeatable, but first we need to get the numbers. We will start by considering the simple stopwatch technique since it can also be performed using many different timing devices from watches to clocks to smart phones and only requires simple division to get an average time per cycle. But, before we continue, we should discuss some of the issues to be considered when selecting start/stop read points when using a stopwatch.

"WE ARE ONLY HUMAN AFTER ALL, RIGHT?"

Some Tips on Choosing Start/Stop Points

The fundamental problem with trying to observe and measure work time in a cell is that the work is continuous in nature and does not stop—which does not allow us very much flexibility. Unfortunately, if we realize that we have made an error, we must stop and take the observation back to the start. Sometimes, making a mistake is unavoidable, but, with some careful thinking and preplanning, we can take steps to minimize the likelihood of error happening.

"SOME TIPS ON CHOOSING START/STOP POINTS."

The first item to consider is exactly what constitutes the beginning of a particular event. Since we will be observing, it seems only natural that we look for visual cues that recur each time that this point in the cycle is reached. For example, this could be a touch as the worker reaches for something. Another possibility is watching for the worker to begin taking a step toward something. Yet, another would be to watch for a light or a light-emitting diode that corresponds to the read point being observed to change states. What matters most is that the observer is able to discern these points every time they occur. Next, we consider whether other cues might work. Sometimes, it is possible to align the read point with a sound instead. The sound could be a click from some kind of a mechanism, the operation of a solenoid, or any other sound that occurs every time the same point is reached. The next consideration is that of something that can be felt. Some things can actually be felt when they occur, especially when larger machines are involved. Regardless of the cues selected, it is critical to be able to consistently recognize them in order to minimize the error from reaction time as well as reduce the opportunities for mistakes. We can now continue with our discussion.

Simple Stopwatch Method

Since we know that the simple stopwatch can only measure a single event, we will measure the time that is taken for some number of consecutive cycles and then divide that time by the total number of complete cycles during that time period. The important aspects are the number of cycles in the average calculation, taking care to avoid including time for partial cycles at the starting and ending points. In the latter case, we can surmise that the points we want to use to synchronize our stop and start actions on the measuring device are when the first cycle in the measurement event begins and when the last full cycle ends. For a single event, the starting and stopping read points are the same; if they are not, then we have introduced some error into the calculation. This particular type of error gets smaller as the number of cycles in the measurement event increases. However, this makes testing for small changes in time due to continuous improvement efforts more time consuming since a large number of cycles may be required based on the amount of potential change. But, the technique itself is quite simple.

CHOOSE THE START/STOP POINTS

First, we must choose a good place in the process to use as the start/stop read point for actuating the button that starts the accumulation of time. As discussed in the "Some Tips on Choosing Start/Stop Points" section, it is very important to select a consistent and repeatable cue that occurs each time this point in the process is reached. If we have good cue, the error introduced by a delay in our physical reaction time will be much less.

BEGIN TIMING WHEN CYCLE STARTS

Once we have selected a good start/stop point cue, we are ready to begin the timing of the observation. Start the timing on the stopwatch by pushing the start/stop button. (Some simple stopwatches may have different names for the buttons.)

KEEP COUNT OF COMPLETED CYCLES

While the stopwatch continues to accumulate the passage of time, keep count of the number of completed cycles. Some people will use a pencil and paper, whereas others may prefer different methods. The key here is to make sure that we get an accurate count. If we use a very large number of completed cycles, the error in calculating the average will not be that significant, but, the smaller the total, the larger the effect the error will have on the average.

"STOP TIMING WHEN STOP POINT OCCURS

When we are ready to stop timing the observation, we must align our stopping point to ensure that the last cycle captured is complete. In most cases, this means that worker may have to walk back from the last station in the process to return to the first station and then reach the point that would signify the beginning of the next cycle. Once this start/stop read point is reached, the timing on the stopwatch is stopped by pressing the appropriate button on your model of stopwatch.

RECORD ACCUMULATED TIME AND COMPLETED CYCLES

After the timing is complete, it is usually a good idea to write down or otherwise record both the total accumulated time and the number of completed cycles. It is especially important to keep the information of the sample observation if we want to understand the history of improvement activity for a particular work cell. Some people do not feel that this is a critical aspect of improvement. However, others like to track not only successful improvement efforts but also those that were not successful.

DETERMINE AVERAGE TIME PER CYCLE

Divide the total accumulated time by the number of full cycles that are completed. This will give us the average time per cycle for this observation. As mentioned earlier in this section, the larger the number of completed cycles in the observation, the lesser impact any error introduced will have on the average time.

Now that we better understand this method, we can summarize the steps, as shown in the following:

1. Choose the start/stop read point for measurement.
2. Start timing on the stopwatch when this point occurs in the work cycle.
3. Keep count of completed cycles while time accumulates on the stopwatch.
4. When stop point occurs on completion of last cycle, stop timing on the stopwatch.
5. Record the total accumulated time and number of completed cycles.
6. Divide the time by the number of cycles to get the average time per cycle.

In this method, we have calculated an average cycle time. How does this average time compare to the TT (customer requirements)? Up to this point, we have been considering cells with only one worker. Some cells will have multiple workers, all trying to get their work done in less than the TT. The assumption is that each worker has the same work time to accomplish; otherwise, some of the workers will be waiting (a form of waste). Another way of looking at this is that the total amount of worker time in a multiple-worker cell should add up to a multiple of TT (e.g., 2 × TT for a two-person cell) if all the workers are fully utilized.

We now understand that the average time for a cycle will most likely vary from day to day, worker to worker, or even for the same worker on different days. It is important to understand that the average, though useful for some purposes such as getting a ballpark estimate for use in comparing as a standard, will not be able to tell us some of the other things that we may want to know. Some of these things include the variation, the *best* and *worst* times, and so on. There are some other issues with this scenario as well. First, how likely is it that the amount of work for each worker will match perfectly to the TT? It is probably not too likely, so, in cells with multiple workers, we can expect mismatches between them, which will result in workers with shorter times waiting on those with longer times. This can sometimes cause the shorter-time worker to adjust his or her pace to try and better match the pace of his or her longer-time counterpart. This a common issue because waiting can appear to be goofing off to those who do not know that the waiting is built-in. In this situation, it can be very difficult to improve since there is also a balancing aspect to multiple-worker cells. So, in order to try and balance the work better between the workers and the machines, we need to be able to break down the work components so that the job can be analyzed for opportunities to transfer the components between all the workers. This is where the memory stopwatch technique can help.

Memory Stopwatch Method #1

The memory stopwatch methods utilize a stopwatch feature that allows single time intervals from an ongoing event to be stored on demand, while the overall time continues to accumulate. These stored readings are often referred to as lap memories since this stopwatch feature is prevalent in sports where individual lap times are of interest. The feature allows us to capture the time for sequential events such as successive laps, while the time for the entire event continues to accumulate. There are several ways that this can be utilized for our purposes. First, the memory stopwatch can be used in the same manner as the simple stopwatch if desired—to get an average time per cycle. However, a more powerful use is to measure the individual times for some number of consecutive cycles. This allows us to compare the variation between cycles by a worker or even compare the variation between different workers doing the same job. This is very useful for pursuing continuous improvement since it can help identify the contributors to variation.

The first method we will discuss using the memory stopwatch is aimed at capturing consecutive individual cycles. The method starts out the same as the simple stopwatch method. Choose the start/stop read point for a complete cycle. Next, while observing the worker performing the work, press the start button when the first cycle begins, and then press the lap button each time that the worker starts another new cycle—except for the last cycle. For the last cycle, we will press the stop button at the point where another new cycle would be starting in order to ensure that the stopping point coincides with the starting point. The method is explained in more detail as follows.

CHOOSE START/STOP POINT

As we did in the simple stopwatch method, we must choose a good place in the process to use as the start/stop read point for actuating the button that starts the accumulation of time. We now understand that it is very important to select a consistent and repeatable cue that occurs each time that this point in the process is reached. If we can find a good cue to use, we can greatly reduce any error that is introduced due to the physical reaction time.

BEGIN TIMING WHEN CYCLE STARTS

In this method, after we have selected a good start/stop point cue, we can begin the timing of the observation. We start the timing on the memory stopwatch by pushing the appropriate button. Several models that we have used over the years have a start/stop button. Your individual model of memory stopwatch may have a different button name, but the action of starting the timing is the desired result.

ACTUATE SPLIT/LAP BUTTON FOR EVERY CYCLE EXCEPT LAST ONE

While the stopwatch continues to accumulate the passage of time, observe for the recurrence of the start/stop read point, which indicates the completion of another cycle. However, with a memory stopwatch, we will actuate the appropriate button that stores the time since the last button actuation in memory. In most models, we have seen that this button was referred to as lap, reset, split, and so on. Regardless of the name, the action desired is to store the last time interval into memory, while the total accumulation of time continues.

STOP TIMING WHEN LAST CYCLE IS COMPLETE

Just as with the simple stopwatch method, when we are ready to stop timing, we must align our stopping point to ensure that the last cycle captured is complete. Once this start/stop read point is reached, the timing on the stopwatch is stopped by pressing the appropriate button on your model of stopwatch. The last time interval is also stored into one of the memories.

RECALL AND RECORD THE CYCLE TIMES

When the timing is complete, it is necessary that we write down or otherwise record memories. In this method, the contents of each memory correspond to the time for a complete cycle, stored in sequence as observed. As discussed in the "Simple Stopwatch Method" section, it is often a good idea to record the information from our improvement efforts. It is also a good idea to do so for this method in order to ensure that we do not accidently clear the memories and lose our information before we are finished with it. Depending on the number of cycles recorded in the observation, you may want to prepare a simple paper form to help organize the data better when we recall them from the stopwatch. Note also that the order that we recall the memories is also the order that the cycles occurred. This can also be useful information as well. The simple paper form can prove to be very valuable in our efforts.

In some instances, a simple table can be sketched out quickly to hold the information. This will suffice in some cases, but it can be difficult for other people to follow. Things can also change over time as we go through our analysis of an observation. Many times, we will think of another aspect that we may be interested in. For example, we are usually interested in using this method in order to capture the individual cycle times. But, once we have these times, we are also interested in looking at the maximum variation. Spreadsheet programs are an easy way to quickly create a prepared form that can not only help us organize our data but also add or change them very quickly. This is not the case for a sketched form, as shown in Figures 4.3 and 4.4.

In this simple prepared form, we can also describe what it is that we are measuring and record the starting and stopping points that are used, the 10 different observed cycle times (OCTs), the total build time for the number of observations, and the average build time per observation. The lower portion helps us select the highest and lowest cycle times (CT_{high} and CT_{low}) so that the maximum variation (V) can be calculated for the set of observed cycles. We are not big fans of forms, but we do like to organize our data in order to better analyze them. The one above can be put together in just a few minutes using a common spreadsheet program, or you can sketch it by hand and make copies.

We can summarize the memory stopwatch method #1 into five steps as follows:

1. Choose a start/stop point for measurement.
2. Start timing on the stopwatch when this point occurs in the work cycle.
3. Actuate the split/lap button for every complete cycle except for the last one.
4. When the stop point occurs on completion of the last cycle, stop timing on the stopwatch.
5. Using a prepared form or just paper, recall and record the cycle times; these are in same order as observed.

Figure 4.3 Hand-sketched prepared form.

		Read points		Observations										Build time total	Average build time
#	Description	Starting	Ending	1	2	3	4	5	6	7	8	9	10		

$$CT_{high} = \underline{\quad} \; ; CT_{low} = \underline{\quad}$$

$$\text{Variation} = \underline{\quad} - \underline{\quad} = \underline{\quad} \; (V)$$

Figure 4.4 Spreadsheet form with additional information.

Variation

In the simple spreadsheet form shown in Figure 4.4, the bottom section introduced a method for determining the variation (*V*) between the highest and lowest values that are observed. This is important in order to better analyze the opportunity for improvement. There will always be some variation, but, if we monitor it closely, we can determine if it is decreasing or increasing. Of course, we would like for it to decrease to as low a value as possible, but we need a way of visually reflecting variation as numbers can sometimes be deceiving based on the magnitude of the observed values in relation to the variation. Figure 4.5 shows a common way to simply reflect variation.

The variation from the best time to the highest or longest time by itself does not show the entire story. Therefore, often, you will see the variation that was reflected along with the actual cycle times that were observed and recorded. This gives us a better perspective of the variation and its magnitude compared to the observations that were measured. An example is shown in Figure 4.6 for a sample of 10 consecutive observations. Notice in this example that the worker is only within the TT for 4 of the 10 cycles that were observed, so there is much improvement to be done. This graph also shows the standard time, whether the TT or the DCT. Sometimes, the graph may also be useful to show the average cycle time as well.

When referring to the best time in this context, this usually refers to the lowest time that was observed and was not considered an anomaly of some sort. An anomaly might be that the reaction time of the person doing the time measurement hit the button on the stopwatch too early or too late, and this caused a reading that was too low. It is pretty obvious that some error or variation can be introduced since we are depending on human senses and reaction times to take the time measurements. If, for some reason, the person taking the measurements reacts too quickly for that read point, it will capture too short of a time.

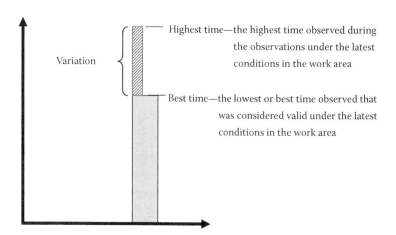

Figure 4.5   Variation is the difference between the highest and the lowest cycle times.

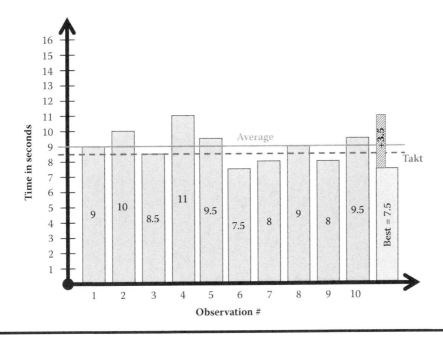

Figure 4.6   Variation is often shown on a graph of observed cycle times.

However, if he or she reacts too slowly to the read point preceding the read point under question, it will also have the same result. Therefore, we must strive to minimize the variation that is introduced due to reaction time by the person taking the time measurements.

As we have already discussed earlier in this section, the reaction time variation can be reduced somewhat by applying the concept of standardization—for example, we can try to find a cue that corresponds to the start/stop point as described earlier, but that is very distinctive and hard to miss each time that it occurs. This increases in importance the smaller the time increments that are measured become; therefore, it is vital that we document the way that we do our observation and measurements so that they can be duplicated in the future. Also, note that this method requires us to record the times that were recalled from memory. Although it is not absolutely necessary, we prefer to use a simple prepared table to record the information so that our information is organized and easily understood by others. If the prepared table is organized properly, recalling the data is very easy. However, as the number of data points increases, the opportunity to make an error when recalling the memories also increases. Also, as the length of the time intervals becomes shorter, the opportunity for mistakes becomes higher. It is for this reason that we prefer to make a prepared form before we start an observation. It causes us to think deeper about what we are doing so that we can set up the form correctly, which then helps us keep the data organized when we recall them from the stopwatch memories. This next method is actually a variation of the one that we have just been discussing with the exception that there are more data points to capture.

Memory Stopwatch Method #2

There are other ways that the memory stopwatch can be used that are of interest to us. We know that by using this type of stopwatch, we can measure consecutive time intervals for an observed event, but we are not limited to just full cycles (or laps)—we can measure individual work components at the stations as well. We are only limited by the number of memories that are available in the stopwatch and our ability to recognize and react to the various start/stop points that separate all the work components. Total cycle times are still available for comparison to the DCT or TT, and the totals can also be used in the same manner as the previous method. However, as we continue our search for ways to improve our operation, we start to look even deeper and find that the next level down from individual full cycles is made up of the work times at each station and the walk times between them. See Figure 4.7.

This method, although very powerful, is a bit more complicated. In the previous method, all that was required was that we watch for a single event—the single start/stop point that occurred once each cycle. In this method, there are several start/stop points, and they occur multiple times each cycle. Notice that the stop point of one component is the start point of the next successive one. In our truck example, recall that we have five work components (for the five stations); then there are five walk components—to get from one station to the next until completion when we return to the first station. So, instead of one time for a full cycle, there are 10 component times. At this point, we start to see the concern for the number of lap memories that are available as they will determine the maximum number of full cycles that this technique can accommodate.

We see now that there are 10 start/stop read points for each full cycle that is captured and that it is necessary to define these points so that we can recognize them easily when they occur. As described in the "Variation" section, the shorter the time intervals being measured, the more distinct the cues must be to improve reaction time. Also, note that since the start point of the next component is the stop point of the previous one, then the stop point of the very last component

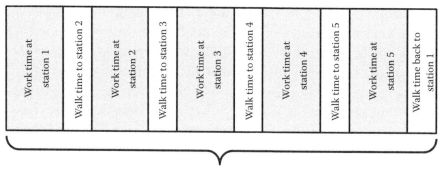

Total cycle time is equal to sum of all the work components.

Figure 4.7   A complete cycle is made up of all the work components.

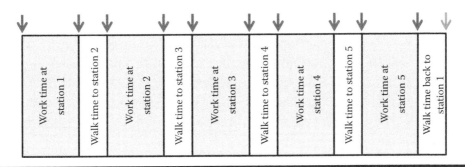

Figure 4.8 Component start/stop points—return to start to complete last cycle.

still includes returning to the start point and actuating the stop button on the stopwatch instead of the lap button (as if another cycle was starting). One way to quickly understand this is to consider a single complete cycle—the last component is not complete until the worker walks back to station 1 and performs an action as if starting the next part, but we stop the accumulation of time rather than use the lap button. See Figure 4.8.

This method can be described as follows.

CHOOSE START/STOP POINTS

As we did in the memory stopwatch method #1, we must choose a good place in the process to use as the start/stop read point for actuating the button that starts the accumulation of time. In both of the previous methods, the start/stop read points were the same place in the cycle. However, in this technique, there are many different points to be chosen. This makes things a bit more complicated, and choosing the correct cues can prove difficult. Often, this method takes a few practice runs in order to get things situated properly.

BEGIN TIMING WHEN CYCLE STARTS

After we have selected the proper cues for our read points, we can begin the timing of the observation. We start the timing on the memory stopwatch in the same way as we do in the memory stopwatch method #1.

ACTUATE SPLIT/LAP BUTTON FOR EVERY CYCLE EXCEPT LAST ONE

In this method, while the stopwatch continues to accumulate time, observe for each successive read point. Remember that there are many different read points and cues to be on the alert for, and therefore it is important to avoid distractions while observing the work. Each time we observe the cue of a read point, we actuate the appropriate button that stores the time interval into memory. This step is similar to the memory stopwatch method #1 with the exception that there are more read points to capture in memory.

STOP TIMING WHEN LAST CYCLE IS COMPLETE

As with both the previous methods, when we are ready to stop timing, we must align our stopping point to ensure that the last cycle (or read point) is captured. Once this read point is reached, the timing on the stopwatch is stopped by pressing the appropriate button on your model of stopwatch. The last time interval is also stored into one of the memories. If we accidentally press the lap button again, the time interval is captured, but the total time continues to accumulate.

RECALL AND RECORD THE CYCLE TIMES

When the timing is complete, we recall and record the memories in a prepared form that is designed for this purpose. In this method, the contents of each memory correspond to the time for a particular work component, stored in the sequence that is observed. We believe that it is a good idea to record the information from our improvement efforts so that we can understand the conditions of the observation as well as the techniques that are used. And, it is always a good idea to do so for this method in order to ensure that we do not accidentally clear the memories before we are finished. Therefore, a prepared form is almost a necessity for this method.

We can summarize the memory stopwatch method #2 into five steps as follows:

1. Choose the read points for measurement.
2. Start timing on the stopwatch at the correct start point in the work cycle.
3. Actuate the split/lap button for every read point except for the last one.
4. When the stop point occurs on completion of the last cycle, stop timing on the stopwatch.
5. Recall and record the work component times in a prepared form.

This last step in this method brings up our next issue for discussion. In our simple stopwatch method, we only had to recognize two points in the timed event to actuate the buttons on the stopwatch—the beginning of the first cycle and the very end of the last one. In the memory stopwatch method #1, we had to recognize multiple points: the beginning of the first cycle and then one point for each full cycle in the observation sample. In the memory stopwatch method #2, we still start with the beginning point of the first cycle, but this time we must recognize multiple read points during each complete cycle before finally ending with the completion of the last cycle (or read point). In our ongoing truck example, if we want to measure 10 complete cycles, this means that we will need a stopwatch with at least 100 memories (10 cycles × 10 components for each cycle). And, once we are finished capturing the work component times in our observation sample, we will need to retrieve the time components in the correct order so that we can analyze the results. This is a really good place for a prepared form. The form can help us to keep our data organized accurately—there are 100 different time measurements in this case. However, at this point, we can start to see the advantages of a prepared form from a spreadsheet over a hand-sketched one. Consider Figures 4.9 and 4.10.

In this more complex prepared form, we can still describe what it is we are measuring, but, now, defining the starting and stopping read points becomes critical since we will be watching for many more different events to occur. Our reaction time is going to be very important as well since the read points will be occurring much more frequently than in our previous methods. Also, note that as the data are recalled, it is crucial that they are entered into the correct location in the observation section. Otherwise, the component times will be in the wrong place, and any analysis would be useless. This is why a prepared form can be very helpful. Upon deeper reflection, we realize that the data in this case will produce a 10-by-10 array. How the prepared form is constructed determines

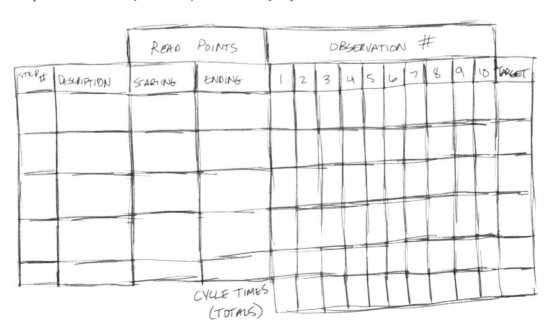

Figure 4.9   Hand-sketched prepared form can get very complex and is difficult to change.

#	Description	Read points		Observations										Row total	Row average	Lowest time
		Starting	Ending	1	2	3	4	5	6	7	8	9	10			
1																
2																
3																
4																
5																
6																
7																
8																
9																
10																
			Total each column to get OCT													

$CT_{high} = \underline{\quad}$; $CT_{low} = \underline{\quad}$

Variation $= \underline{\quad} - \underline{\quad} = \underline{\quad}$ (V)

Figure 4.10 Spreadsheet form is easy to change and can be expanded.

how the data should be entered as they are recalled from the memories in the stopwatch. In our example form, we have set it up to enter the data starting at the top of the first column in the observation section and then enter the data point into the box immediately below it until we reach row 10. Once the box on row 10 is used, we begin again on the first box of the next column, as shown in Figure 4.11. If we did everything correctly, we will finish with the box on row 10 column 10. If we did not, we should start the recall process over with either a new form or after we have erased the data in the observation section. For this reason, we highly recommend using a pencil with an eraser.

Notice on our prepared form that each row corresponds to a work component for a complete cycle. Figure 4.11 also shows that there are places for the totals

#	Description	Read points		Observations										Row total	Row average	Lowest time
		Starting	Ending	1	2	3	4	5	6	7	8	9	10			
1																
2																
3																
4																
5																
6																
7																
8																
9																
10																
			Total each column to get OCT													

$CT_{high} = \underline{\quad}$; $CT_{low} = \underline{\quad}$

Variation $= \underline{\quad} - \underline{\quad} = \underline{\quad}$ (V)

Figure 4.11 It is important to understand how the data are entered into the form.

of both rows and columns. If we enter the total of the data that were entered in each row in the column called "Total," we can then calculate an average for that component by dividing the total by the number of observations that were entered—in our example, this is 10. If we then total the data that were entered in each column in the boxes below row 10, we get the total time for each complete cycle. Note that we can also get a total of the average times for each component. Some people like to use this as a DCT value.

"SOMETIMES IT IS A BIG DEAL TO HAVE TO START OVER!"

Summary of the Three Stopwatch Methods

We have gone through the three main stopwatch methods rather quickly. This is probably a good point to compare them together in order to more easily understand the basic differences. Figure 4.12 illustrates the main similarities and differences between the three methods.

Notice that all three use the same points to start and stop the total accumulation of time during an observation. In the simple stopwatch method, we can only calculate the average time per cycle. In some cases, this is sufficient. In the memory stopwatch method #1, we use the lap memory feature of the stopwatch in order to capture the individual cycle times during the observation. According to the number of memories involved, it is often useful to utilize a prepared form for recording the data upon recall. And, finally, in the memory stopwatch method

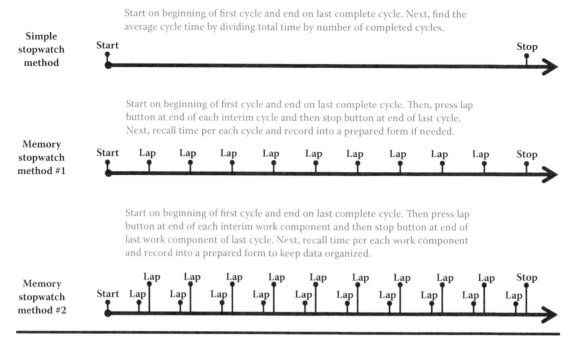

Figure 4.12 Comparison of the three stopwatch methods.

#2, we use the lap memory feature a little differently in order to capture the work component times instead of just the individual cycle times. This method almost always requires a prepared form in order to recall and record the data in the proper sequence. Once the data are recalled, the components can still be added together in order to determine the individual cycle times.

Food for Thought

In the "Stopwatches" section, we mentioned a method that is called *lowest repeatable*. Some people prefer to use the lowest entry for a component that appears to repeat in the data that were gathered in the observations. However, sometimes, the data do not repeat within the small number of particular work components that we are able to accumulate due to the number of lap memories that are available. Even if we use a stopwatch with several hundred lap memories, gathering and keeping this amount of data organized can get extremely tedious and time consuming—a sometimes seemingly impossible feat.

However, there are other ways that we have seen that are used for choosing numbers to utilize in situations like this. Some will just select the lowest time that is observed. The idea is that this is the best time that is observed and represents a target that has been demonstrated at least once before. We do not prefer this method for fairly short work component times since errors in reaction times can greatly impact values in this technique. If the numbers are larger, slight variations in reaction time to activate the stopwatch buttons have much less of an impact on the values, and this technique can be quite reliable. If we were going to use this technique with the prepared form in our previous examples, we could simply circle the best time in each row, and the sum of these would represent a theoretical target that is made up of the best times for each component. However, we must bear in mind that a technique like this yields a total that has not been demonstrated in real life. It is just a way of coming up with a theoretical target time.

"SOME PROBLEMS CAN SEEM OVERWHELMING."

There are probably many more techniques that can be used to try and determine a suitable DCT. However, we do not prefer to use the more elaborate or complex techniques. In most instances, we have found that the simple techniques will usually suffice. TT is the target that is of interest for day-to-day operation. The DCT is something that we prefer to use for continuous improvement purposes to represent our baseline for making improvements on individual worker jobs or workstations. It is this latter case that leads to one last technique that we would like to discuss.

As we learn to think deeper about the various work components that make up the worker's job in a work cell, we begin to see that it is often necessary to select a particular workstation upon which to focus our improvement efforts. The second lap stopwatch method can help us with this as well if the component times are not too short for our reaction time. Suppose that we wanted to break down the work components of an individual workstation in a work cell. We would still identify the appropriate start/stop read points like before, with two main ways of approach. In either approach, we need to recognize that we are not interested in the times for the other stations. In the first approach, we will need a stopwatch that has a feature to stop accumulating time until the worker returns to the station where the timing can continue again—a task that can be extremely difficult and prone to errors. The second approach is much easier; we can simply allow time to accumulate between successive instances of the worker returning to the selected station and have our prepared form account for an accumulated time so that this time can be recorded on the prepared form but then filtered out of our analysis.

Some Other Information on the SWC

Before we move on from our discussion on the SWC, it is important to note some other information that is often found on it. We know that there is a visual aspect to the SWC as the layout sketch is the central focus of this standardized work tool. Another visual component that is often found here are some simple symbols. These symbols denote things such as safety issues for the worker, quality checkpoints, and places where parts should be (or sometimes should not be) during the cycle. Figure 4.13 shows some typical symbols.

Other information sometimes found on an SWC includes things like the name of the operation or job, the last revision information, the person who created it and/or who last modified it, and other information that is useful for the

Figure 4.13   Some common symbols used on an SWC.

organization. We even prefer to list all the tasks somewhere on the SWC as well since, in real life, the worker is often required to perform some tasks on demand as needed in addition to the cell work. We discussed this at some length in *New Horizons in Standardized Work* (Martin and Bell 2011, pp. 37–46).

"WE NOW SAY GOODBYE TO STOPWATCHES AND THE SWC!"

Chapter 5

Questions for Standardized Work Chart Review

Introduction

"PROBLEMS JUST KEEP ON COMING."

In Chapter 4, we learned about the standardized work chart (SWC) and how it can help us better understand the relationship between the geographic layout of the process steps and the impact that this can have upon the various work components as well as our customer's requirements. During this time, it was also necessary to learn about how a standard time, whether takt time (TT) or desired cycle time (DCT), was necessary to use as a basis for comparison in order to determine if we were indeed making progress in our continuous improvement efforts. It is entirely possible that some of our improvement ideas could actually make things worse, so the standard time is used to serve as a point of reference to not only reduce variation but also serve as our basis for evaluating whether we are improving or not. We also discovered that in order to measure the observed cycle times (OCTs), the two basic types of stopwatches both had methods that would allow us to use their capabilities for our improvement efforts.

However, to effectively use the two basic types of stopwatches, it was necessary that we understand how to break the work cycles into measurable time components. In some cases, this was simply complete cycles, whereas, in others, it was various components of the work cycle—whether it was work, walk, wait, or machine time. We also learned that variation was another issue that arises once we start looking deeper into the various work components. And, finally, we learned that in order to reduce the chances for error, it is extremely helpful to plan ahead and prepare a table or form to help keep the numerous measurements organized for accurate analysis. Next, we will work through some exercises and questions based on the material in Chapter 4.

Question 1: Calculate the TT for a customer demand of 150 parts per day.

Given

1. Production is scheduled for one shift per day.
2. The workday is 8 hours.
3. There is a 10-minute huddle meeting at the beginning of the shift.
4. There are two 10-minute breaks per shift: one in the morning and one in the afternoon.

Remember: 1 hour = 60 minutes

Use the answer area for question 1 for your answer.

ANSWER AREA FOR QUESTION 1

It is possible that we could be presented with a different version of the above problem. In some cases, we may have a work cell that produces different models of a product or even many different products. This is a very common occurrence in low-volume but high-mix scenarios. The processes may be the same, but the work content varies based on the process characteristics such as labor that is required, machine time, the number of components, and so on. When a new model or version is introduced, it is sometimes necessary to work backwards to determine the TT or the DCT based on the amount of work content that was estimated in the design phase or measured in a prototype environment. Regardless of how the work content was determined or estimated, it does occur quite often, and we should be prepared to deal with a situation that is similar to the following question.

"SOME PROBLEMS HAVE TO BE LOOKED AT FROM A DIFFERENT ANGLE."

Question 2: To produce a part for the customer, it takes 10 minutes of total work time. If the customer TT is 3 minutes per part, what would be the required minimum number of workers to meet the customer takt?

Hint: Round up to the nearest number of workers.

Use the answer area for question 2 for your answer.

ANSWER AREA FOR QUESTION 2

Question 3: (True or false) In referring to question 2, all four workers do not have to have a DCT that is equal to or less than 3 minutes per part to complete to meet the customer demand.

Use the answer area for question 3 for your answer.

```




```

ANSWER AREA FOR QUESTION 3

Question 4: (True or false) The read points for establishing the cycle beginning and cycle ending can be visual cues as well as sound cues.

Use the answer area for question 4 for your answer.

```


```

ANSWER AREA FOR QUESTION 4

"GOOD CUES ARE IMPORTANT TO OUR OBSERVATIONS."

Question 5: In using the simple stopwatch method, you have observed 30 cycles for a total of 900 seconds. What is the average cycle time in seconds?

Use the answer area for question 5 for your answer.

ANSWER AREA FOR QUESTION 5

Question 6: (True or false) It is very important to accurately record the number of observations that were taken over the accumulated time of a simple stopwatch study.

Use the answer area for question 6 for your answer.

ANSWER AREA FOR QUESTION 6

Question 7: For a memory stopwatch, how many memories would be required to capture 10 individual observed cycles?

Use the answer area for question 7 for your answer.

ANSWER AREA FOR QUESTION 7

Question 8: (True or false) The memory stopwatch method #1 can more accurately capture variation from cycle to cycle as compared to the simple stopwatch method.

Use the answer area for question 8 for your answer.

ANSWER AREA FOR QUESTION 8

"WE CANNOT AFFORD TO BE DISTRACTED DURING OUR OBSERVATIONS."

Question 9: How many full cycles can we capture from the example in shown in the work component breakdown for question 9 using the memory stopwatch method #2 with a stopwatch that has 100 memories?

Use the answer area for question 9 for your answer.

Work time at station 1	Walk time to station 2	Work time at station 2	Walk time to station 3	Work time at station 3	Walk time to station 4	Work time at station 4	Walk time to station 5	Work time at station 5	Walk time back to station 1

Total cycle time is equal to sum of all the work components

WORK COMPONENT BREAKDOWN FOR QUESTION 9

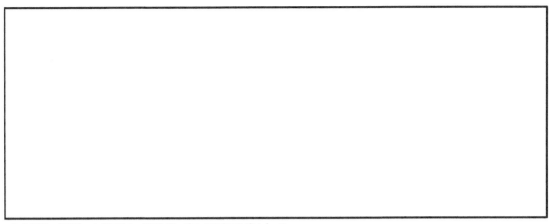

ANSWER AREA FOR QUESTION 9

Question 10a: Complete the example observation form for question 10a.

#	Description	Read Points Starting	Read Points Ending	1	2	3	4	5	6	7	8	9	10	Total	Avg.
1	Work Time at Station 1	Reach for Truck Body	Turn towards Station 2	4.2	3.8	3.6	4.2	3.9	3.8	4.3	3.6	3.6	4.2	39.2	
2	Walk time to Station 2	Turn towards Station 2	Stop at Station 2	2.5	2.4	2.5	2.6	2.5	2.4	2.5	2.4	2.4	2.5	24.7	
3	Work Time at Station 2	Stop at Station 2	Turn towards Station 3	5.5	4.3	5.7	4.2	4.5	5.0	4.3	4.3	5.2	5.5	48.5	
4	Walk time to Station 3	Turn towards Station 3	Stop at Station 3	2.2	2.1	2.0	2.0	2.0	2.2	2.2	2.1	2.3	1.8	20.9	
5	Work Time at Station 3	Stop at Station 3	Turn towards Station 4	5.0	6.5	5.5	5.3	5.0	5.2	6.0	5.5	5.4	5.0	54.4	
6	Walk time to Station 4	Turn towards Station 4	Stop at Station 4	2.3	2.1	2.0	2.3	2.2	2.0	2.1	1.9	2.0	2.1	21.0	
7	Work Time at Station 4	Stop at Station 4	Turn towards Station 5	5.1	4.7	4.4	4.4	4.8	4.9	4.4	5.0	4.8	5.2	47.7	
8	Walk time to Station 5	Turn towards Station 5	Stop at Station 5	2.4	2.5	2.4	2.3	2.4	2.5	2.4	2.5	2.5	2.4	24.3	
9	Work Time at Station 5	Stop at Station 5	Turn towards Station 1	5.6	5.5	5.0	5.5	5.5	5.0	6.0	5.4	5.0	5.5	54.0	
10	Walk time to Station 1	Turn towards Station 1	Reach for Truck Body	3.0	3.1	3.2	2.9	3.0	3.1	3.2	3.0	3.1	3.0	30.6	
			Add all to get OCT:	37.8	37.0	36.3	35.7	35.8	36.1	37.4	35.7	36.3	37.2	365.3	

$$CT_{high} = \_\_\_\_ \; ; \; CT_{low} = \_\_\_\_$$

$$\text{Variation} = \_\_\_\_ - \_\_\_\_ = \_\_\_\_ \; (V)$$

EXAMPLE OBSERVATION FORM FOR QUESTION 10a

Question 10b: Where are the best opportunities to improve the work components?

Hint: Compare work component averages to the lowest repeatable times.

Use the answer area for question 10b for your answer.

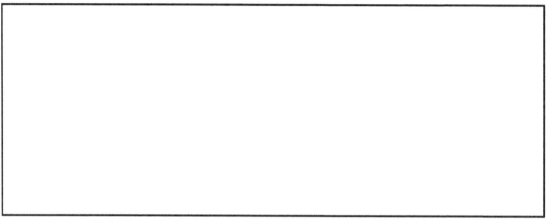

ANSWER AREA FOR QUESTION 10b

Question 10c: Will your opportunities for improvement meet the TT shown on the example SWC for question 10c?

Use the answer area for question 10c for your answer.

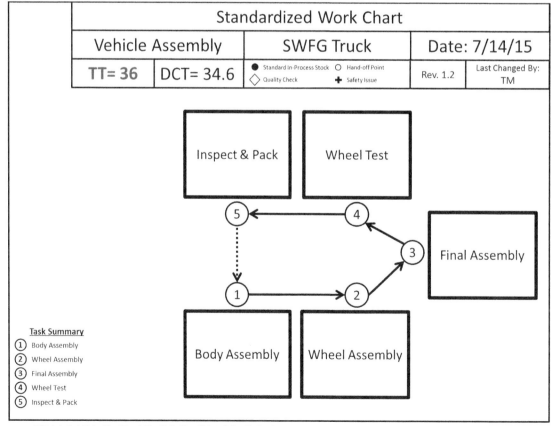

EXAMPLE SWC FOR QUESTION 10c

ANSWER AREA FOR QUESTION 10c

Question 10d: Draw a bar graph reflecting the individual cycles, the best time that was observed as a target with the variation that is shown, and, finally, add a line that shows the average cycle time using the values from the table in this example.

Hint: The *best* time is the lowest OCT in this case.

Use the answer area for question 10d for your answer.

ANSWER AREA FOR QUESTION 10d

"SOME PROBLEMS MUST BE BROKEN DOWN INTO SMALLER PIECES."

Answers for SWC Questions

"WORKING OUT THE DETAILS MAKES THE SOLUTION SEEM SIMPLER."

Question 1: Calculate the TT for a customer demand of 150 parts per day.

Given

1. Production is scheduled for one shift per day.
2. The workday is 8 hours.
3. There is a 10-minute huddle meeting at the beginning of the shift.
4. There are two 10-minute breaks per shift: one in the morning and one in the afternoon.

Remember: 1 hour = 60 minutes

See answer for question 1.

a) 8 hours per work day multiplied by 60 minutes per hour = 480 minutes per work day
b) 480 minutes per work day minus10 minute huddle meeting = 470 minutes per work day
c) 470 minutes per work day minus (2 multiplied by 10 minute work breaks) = 450 minutes per work day
d) 450 minutes per work day divided by 150 parts per day customer demand = 3 minutes per part Takt Time

ANSWER FOR QUESTION 1

"IT IS IMPORTANT TO TRY OUT OUR IMPROVEMENT IDEAS QUICKLY!"

Question 2: To produce a part for the customer, it takes 10 minutes of total work time. If the customer TT is 3 minutes per part, what would be the required minimum number of workers to meet the customer takt?

Hint: Round up to the nearest number of workers.

See answer for question 2.

> 10 minutes of total work time per part divided by 3 minutes per part Takt Time = 3.33 workers approximately 4 workers

ANSWER FOR QUESTION 2

In this last example, we are faced with a situation that calls for a number of workers that would be difficult to acquire. There are several options that could be pursued, each of which would make for a fairly enlightening exercise. However, the intent is not to learn how to match labor with work but rather how to strive for continuous improvement. This exercise should help us recognize that there may be opportunities for improvement that would reduce the total work time to allow us to eventually get down to three workers.

Question 3: (True or false) In referring to question 2, all four workers do not have to have a DCT that is equal to or less than 3 minutes per part to complete to meet the customer demand.

See answer for question 3.

> **The answer is False.** Each work must have a desired cycle time less than Takt Time or the combination of work and workers will not meet the customer requirement.

ANSWER FOR QUESTION 3

"CONTINUOUS IMPROVEMENT MEANS WE ARE NEVER DONE!"

Question 4: (True or false) The read points for establishing the cycle beginning and cycle ending can be visual cues as well as sound cues.

See answer for question 4.

> **The answer is True.** Visual read points are usually the most common form of detection, but sound cues are effective especially if line of sight is not feasible.

ANSWER FOR QUESTION 4

Question 5: In using the simple stopwatch method, you have observed 30 cycles for a total of 900 seconds. What is the average cycle time in seconds?

See answer for question 5.

> 900 seconds of total work time divided by 30 observed cycles = 30 seconds average per work cycle

ANSWER FOR QUESTION 5

"WE CANNOT AFFORD TO SLOW DOWN ON OUR EFFORTS TO IMPROVE."

Question 6: (True or false) It is very important to accurately record the number of observations that were taken over the accumulated time of a simple stopwatch study.

See answer for question 6.

> **The answer is True.** If we make an error in recording the number of observations taken over the accumulated work time it will result in an error in the average work cycle time.

ANSWER FOR QUESTION 6

In this last question, even though an error would result, the impact of that error is inversely proportional to the total number of cycles in the observation. In other words, the fewer the number of total cycles, the greater the impact the error would have on the average, and the higher the number, the lesser the impact. However, this is not meant to imply that error is acceptable, only that this should be taken into account when analyzing the situation.

Question 7: For a memory stopwatch, how many memories would be required to capture 10 individual observed cycles?

See answer for question 7.

> 10 memories would be required for 10 individual observed cycles.

ANSWER FOR QUESTION 7

"THE SEARCH FOR A BETTER WAY IS NEVER ENDING."

This last question is pretty straightforward; the number of memories in a stopwatch with the lap memory feature determine how many measurements you can store before having to stop and recall the data from the memories. It does not matter whether they are complete cycles or individual work components. It is important to note that when using the simple stopwatch method, there are no memories, and only one time can be captured per observation.

Question 8: (True or false) The memory stopwatch method #1 can more accurately capture variation from cycle to cycle as compared to the simple stopwatch method.

See answer for question 8.

The answer is True.

ANSWER FOR QUESTION 8

The memory stopwatch method #1 will capture each completed work cycle time in an observation allowing us to understand the variation between individual cycles. This is in contrast to the simple stopwatch method, which only allows us to calculate an average cycle time for an observation and will spread any cycle-to-cycle variation across all cycles. There are definite advantages in being able to measure cycle-to-cycle variation. Although identifying some of these advantages would serve as a very good exercise question, our intent is not to justify the memory stopwatch methods but rather to learn to use them in pursuit of continuous improvement.

Question 9: How many full cycles can we capture from the example shown in the work component breakdown for question 9 using the memory stopwatch method #2 with a stopwatch that has 100 memories?

See the read points for the work component breakdown and answer for question 9.

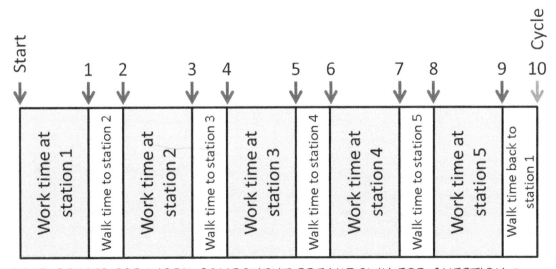

READ POINTS FOR WORK COMPONENT BREAKDOWN FOR QUESTION 9

a) First we determine how many work components there are in a complete cycle by defining the read points for each as indicated in the figure showing **read points for work component breakdown for question 9.**

b) 100 memories divided by 10 work components per total work cycle = 10 individual work cycles.

ANSWER FOR QUESTION 9

Question 10a: Complete the example observation form for question 10a.

See the completed observation form for question 10a (completed values circled in red).

#	Description	Read Points Starting	Read Points Ending	1	2	3	4	5	6	7	8	9	10	Total	Avg.
1	Work Time at Station 1	Reach for Truck Body	Turn towards Station 2	4.2	3.8	3.6	4.2	3.9	3.8	4.3	3.6	3.6	4.2	39.2	3.9
2	Walk time to Station 2	Turn towards Station 2	Stop at Station 2	2.5	2.4	2.5	2.6	2.5	2.4	2.5	2.4	2.4	2.5	24.7	2.5
3	Work Time at Station 2	Stop at Station 2	Turn towards Station 3	5.5	4.3	5.7	4.2	4.5	5.0	4.3	4.3	5.2	5.5	48.5	4.9
4	Walk time to Station 3	Turn towards Station 3	Stop at Station 3	2.2	2.1	2.0	2.0	2.0	2.2	2.2	2.1	2.3	1.8	20.9	2.1
5	Work Time at Station 3	Stop at Station 3	Turn towards Station 4	5.0	6.5	5.5	5.3	5.0	5.2	6.0	5.5	5.4	5.0	54.4	5.4
6	Walk time to Station 4	Turn towards Station 4	Stop at Station 4	2.3	2.1	2.0	2.3	2.2	2.0	2.1	1.9	2.0	2.1	21.0	2.1
7	Work Time at Station 4	Stop at Station 4	Turn towards Station 5	5.1	4.7	4.4	4.4	4.8	4.9	4.4	5.0	4.8	5.2	47.7	4.8
8	Walk time to Station 5	Turn towards Station 5	Stop at Station 5	2.4	2.5	2.4	2.3	2.4	2.5	2.4	2.5	2.5	2.4	24.3	2.4
9	Work Time at Station 5	Stop at Station 5	Turn towards Station 1	5.6	5.5	5.0	5.5	5.5	5.0	6.0	5.4	5.0	5.5	54.0	5.4
10	Walk time to Station 1	Turn towards Station 1	Reach for Truck Body	3.0	3.1	3.2	2.9	3.0	3.1	3.2	3.0	3.1	3.0	30.6	3.1
			Add all to get OCT:	37.8	37.0	36.3	35.7	35.8	36.1	37.4	35.7	36.3	37.2	365.3	36.5

$$CT_{high} = 37.8; \quad CT_{low} = 35.7$$

$$\text{Variation} = 37.8 - 35.7 = 2.1 \quad (V)$$

COMPLETED OBSERVATION FORM FOR QUESTION 10a

Question 10b: Where are the best opportunities to improve the work components?

Hint: Compare work component averages to the lowest repeatable times.

See observation form containing the lowest repeatable values for the work elements (completed values circled in red) and the answer for question 10b.

#	Description	Read Points		Observations										Total	Avg.
		Starting	Ending	1	2	3	4	5	6	7	8	9	10		
1	Work Time at Station 1	Reach for Truck Body	Turn towards Station 2	4.2	3.8	3.6	4.2	3.9	3.8	4.3	3.6	3.6	4.2	39.2	3.9
2	Walk time to Station 2	Turn towards Station 2	Stop at Station 2	2.5	2.4	2.5	2.6	2.5	2.4	2.5	2.4	2.4	2.5	24.7	2.5
3	Work Time at Station 2	Stop at Station 2	Turn towards Station 3	5.5	4.3	5.7	4.2	4.5	5.0	4.3	4.3	5.2	5.5	48.5	4.9
4	Walk time to Station 3	Turn towards Station 3	Stop at Station 3	2.2	2.1	2.0	2.0	2.0	2.2	2.2	2.1	2.3	1.8	20.9	2.1
5	Work Time at Station 3	Stop at Station 3	Turn towards Station 4	5.0	6.5	5.5	5.3	5.0	5.2	6.0	5.5	5.4	5.0	54.4	5.4
6	Walk time to Station 4	Turn towards Station 4	Stop at Station 4	2.3	2.1	2.0	2.3	2.2	2.0	2.1	1.9	2.0	2.1	21.0	2.1
7	Work Time at Station 4	Stop at Station 4	Turn towards Station 5	5.1	4.7	4.4	4.4	4.8	4.9	4.4	5.0	4.8	5.2	47.7	4.8
8	Walk time to Station 5	Turn towards Station 5	Stop at Station 5	2.4	2.5	2.4	2.3	2.4	2.5	2.4	2.5	2.5	2.4	24.3	2.4
9	Work Time at Station 5	Stop at Station 5	Turn towards Station 1	5.6	5.5	5.0	5.5	5.5	5.0	6.0	5.4	5.0	5.5	54.0	5.4
10	Walk time to Station 1	Turn towards Station 1	Reach for Truck Body	3.0	3.1	3.2	2.9	3.0	3.1	3.2	3.0	3.1	3.0	30.6	3.1
			Add all to get OCT:	37.8	37.0	36.3	35.7	35.8	36.1	37.4	35.7	36.3	37.2	365.3	36.5

$$CT_{high}= \underline{37.8}\ ;\ CT_{low}= \underline{35.7}$$

$$Variation = \underline{37.8} - \underline{35.7} = \underline{2.1}\ \ (V)$$

LOWEST REPEATABLE VALUES FOR WORK ELEMENTS FOR QUESTION 10b

a) The variation from the Highest Cycle time to the Lowest Cycle Time is 2.1 seconds, which is predominately seen in the work at each of the Stations.
b) Station 1 has 0.3 seconds, Station 2 has 0.6 seconds, Station 3 has 0.4 seconds, Station 4 has 0.4 seconds and Station 5 has 0.4 seconds of opportunities.

ANSWER FOR QUESTION 10b

In question 10b, we compared the work component averages to the lowest repeatable values that were observed. Sometimes, there may be no clearly repeatable value, so this does not always prove to be as easy as the example table. In such cases, there are other techniques for choosing a target value for each component. If we think deeper and try to develop a list of possible alternative ways, we would have another good exercise question. But, again, this is not the point. What is important is that we might want to define a target value for each of the work components for our improvement efforts, and the value that is selected should be a fairly accurate reflection of the typical time. Some people choose to use the lowest value that is observed for each component as they feel that this represents the best time that was observed. However, this is not without its own issues to consider. Notice that the walk times were not discussed in this question. In this particular example, there was not a significant amount of variation in the walk times. Although there certainly may be an opportunity to improve them as well, the current variation is so small that it could easily be caused by user error operating the stopwatch during observation and recording. This could also be used as an argument against using the best or lowest value unless it had also shown that it is repeatable.

Question 10c: Will your opportunities for improvement meet the TT shown on the example SWC for question 10c?

See answer for question 10c.

a) The average cycle time was observed to be 36.5 seconds.
b) If each work station's average cycle time can be improved to meet the lowest repeatable cycle time consistently, then 2.1 seconds of improvement may be achieved.
c) The average would be reduced to 36.5 seconds minus 2.1 seconds = 34.4 seconds.
d) This would meet the Takt Time (TT) and also meet the Desired Cycle Time (DCT) of 34.6 seconds.

ANSWER FOR QUESTION 10c

Question 10d: Draw a bar graph reflecting the individual cycles, the best time observed as a target with variation that is shown, and, finally, add a line that shows the average cycle time using the values from the table in this example.

Hint: The best time is the lowest OCT in this case.

See graph showing OCTs and variation for question 10d.

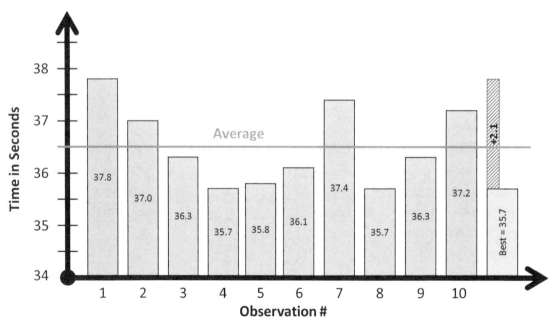

GRAPH SHOWING OCT's AND VARIATION FOR QUESTION 10d

Using simple graphs can help us look for even more opportunities to improve. In this last example, we see that even within 10 consecutive worker cycles, there is over 2 seconds of variation. We know that some variation is inevitable, but that does not mean that we do not continue to try to eliminate it. Striving for perfection, even though we know that we can never completely achieve it, is an important driving force in the continuous improvement mindset. It is one of the things that keep us looking for a better way. In this last example, we also notice that the best time of 35.7 seconds shows up twice in only 10 consecutive cycles, which is an indicator that it might not have been due to an error in measurement. We can continue to look for eliminating the cause of the variation between cycles, even while looking for other improvements such as reduction in the walk time.

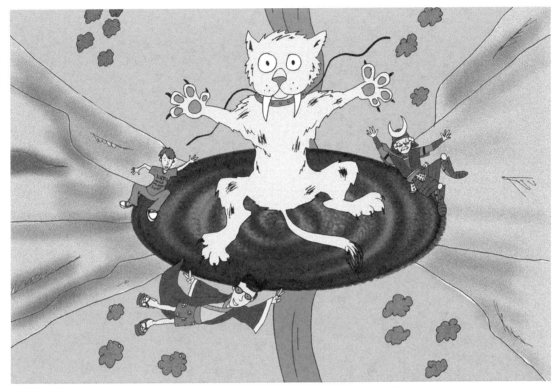

"ON TO THE NEXT ADVENTURE!"

Chapter 6

Work Combination Table: Where Time and (Work) Space Collide

Introduction

"WHERE TIME & (WORK) SPACE COLLIDE!"

We learned in Chapters 4 and 5 about the standardized work chart, where it became clear that we are building upon the idea of continuous improvement. We will continue along this path as we learn about how we can tie all of the work components together and find even more ways to reveal more problems and issues with the worker's job in a work cell.

In this section, we will learn how to develop and use the work combination table (WCT), which allows us to take the work components and put them together in a diagram that helps us see the entire work cycle from beginning to end along with the interaction of any machine cycle times. We now know that after the standardized work has been defined, the next step is to use the standardized work tools to begin the *kaizen* process. When dealing with repetitive tasks involving a combination of work, walk, wait, and/or machine times, a WCT is a very powerful tool for evaluating improvement opportunities. It provides a visual depiction of the relationship between individual tasks in the work sequence, the cycle time of the machines that are involved, and the target standard, usually takt time (TT) or the desired cycle time (DCT).

Although the WCT is a good tool to help identify and quantify opportunities for improvement, it can also be a useful tool when performing a detailed analysis on changeovers, machine cycles, or just manual work sequences in general. It can be used as a tool to analyze any repeatable sequence. Although the name implies that it shows the combination of the work between man and machine, it can be used for improvement efforts of repeatable work elements as well. Examples include product changeovers, delivery routes, or even administrative tasks. The left side of the WCT contains the work elements and times in a table format. The right side contains a graphic representation of those elements as they relate to the geographic layout. In essence, the WCT truly represents a place where time and (work) space collide.

TRANSLATION: "ARE THESE LEAN GUYS FOR REAL OR JUST MORE SPACE CADETS?"

Before we start looking at how to go about creating a WCT, we should pause for a moment and consider that many Lean concepts, principles, and tools are not always obvious or intuitive. There are quite a few people who think that many Lean methods are unorthodox at best and downright detrimental at worst. Hopefully, this chapter will help people better understand how powerful the WCT can be when accurately created and properly used.

In order to create a WCT, we must first observe the process sufficiently to get a good understanding of how the work elements and machine time(s) interact. Once the relationship of the elements is understood, it is necessary to measure all the components next so that the combination of elements can be documented and graphed on the WCT tool, taking care to ensure that all of the problems are shown as well. After an accurate representation is captured, this model can then be used to analyze the process so that opportunities for improvement can be considered. If the problems are not captured and shown properly, it will greatly impact the analysis and the effectiveness of the improvements that are evaluated. Fortunately, the tools and methods we have learned in Chapters 1 through 5 will help provide the information that we need to build a WCT.

"LEAN CAN HELP US LEARN TO REVEAL PROBLEMS AND SOLVE THEM QUICKLY."

"AS WE TRANSITION TO LEAN, SOME PROBLEMS MAY SEEM MORE DIFFICULT."

"DOES LEAN LAND REALLY EXIST–AND HOW DO WE FIND IT?"

How-To: WCT

START WITH A LAYOUT SKETCH.

A WCT begins with the creation of a reasonably accurate layout sketch of the work flow. As we learned in Chapters 2 and 3, in order to create an accurate

Figure 6.1   Sketch the geographic layout of the work area.

layout sketch, it is important to understand the geographical perspective of the equipment and its relationship with the sequence of the work flow. We first draw a sketch that represents the geographic layout of the work area. Next, using a number inside of a circle, label each process step location in the work sequence as it is performed even if the worker must walk back and forth or even retrace steps. Then, use arrows to show movement from one location to another—solid line arrows for the steps that are internal to the sequence and a dashed line arrow to show when the operator has completed an entire cycle and is about to start the next complete cycle. Refer to Figure 6.1.

DRAW A ROUGH DIAGRAM BASED ON THE LAYOUT SKETCH.

Once the layout sketch is complete, the next step is to observe the work as it occurs. During this observation, it is necessary to identify all the relevant work elements so that they can be measured. In most instances, these work elements will be composed of work, walk, and/or wait by the worker, plus any applicable machine time at each work step. It often helps to take a piece of scrap paper and make a rough graph to help visualize the relevant work elements for the process being observed. In our example, the worker performs some work at step 1, then walks to step 2 where some additional work is performed, and walks on to continue the cycle at step 3, while the machine at step 2 performs its automatic cycle and so on through step 5 and then returns to step 1 where the cycle repeats. Refer to Figure 6.2.

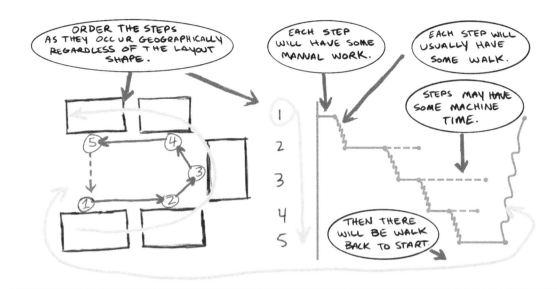

Figure 6.2   Sketch rough graph based on layout.

DEFINE READ POINTS & CAPTURE THE TIMES FROM OBSERVED CYCLES

As we learned in Chapters 4 and 5, once a rough graph is made, we look for the points that define the separation of each work element. Fortunately, the machine time is relatively simple and can be captured separately from the work elements. However, in order to clearly define the start and stop points of each work element, we need some recurring recognizable reference point that our senses can readily distinguish. Once these work elements are captured using the predefined start/stop points, we transfer them from the stopwatch memory to a table (such as the example that is shown in Figure 6.3), usually a simple form that is made for this purpose.

#	Description	Read Points		Observations										Row Total	Row Average	Lowest Time
		Starting	Ending	1	2	3	4	5	6	7	8	9	10			
1																
2																
3																
4																
5																
6																
7																
8																
9																
10																
		Total each column to get Observed Cycle Time (OCT)														

CT_{high}= ___ ; CT_{low}= ___

Variation = ___ - ___ = ___ (V)

Figure 6.3   Identify work element start/stop points and capture times.

Since there will be variation, we must determine which work element times to use—using whatever method seems most appropriate. Totaling these work element times results in the DCT, or what we reasonably expect to achieve each cycle under normal conditions. Recall that the DCT is not the same as the TT, which is based on customer requirements under our chosen operating conditions. If the DCT is not sufficient to meet customer requirements, then steps must be taken to reconcile this problem. For example, the short term might be to add additional work hours, whereas, in the long term, we would definitely be relying on *kaizen.*

ENTER THE CAPTURED DATA INTO THE WCT FORM

After selecting the work element times to use from our measurements, using a WCT form, enter the following work element details into the table:

■ Step number. (The sequence should match the geographic layout of the worker path rather than the work sequence—see the examples that follow.)
■ Work element name or description of the work to be performed at that step.
■ Manual time that the worker spends at the location (as previously selected).
■ Machine cycle time that occurs (where applicable).
■ Walk time to the next step in the work sequence—notice in the table in Figure 6.4 that the walk time is positioned such as to show the beginning and ending steps for that particular walk. (Again, use the times that were previously selected.)
■ Enter the total manual work time and total walk time at the bottom of the table as shown; If there is any forced wait time, enter that at the bottom of the table as well—notice that, if done correctly, the sum of these three values will equal the DCT. (Refer to Figure 6.4.)

CREATE THE WORK COMBINATION DIAGRAM FROM THE DATA IN THE TABLE

Next, sketch the combination of the operator and machine work as a graph using the work element times from the table. Also, show the expected DCT or the TT when applicable, with vertical lines. Walking is usually shown as a wavy line and machine time as a dashed line. Also, note that if there is any forced wait, it is denoted using a double arrow. The comparison of the graph to the left-side data table is shown in Figure 6.5.

It is important to note that there are some cases where the DCT is more relevant than the TT, for example, a very long machine changeover. In the change-over, the TT might not apply, but certainly a standard time would—hence the term desired cycle time. In either case, the standard time should be drawn on the

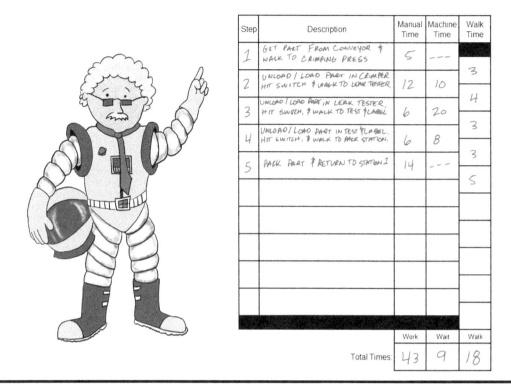

Step	Description	Manual Time	Machine Time	Walk Time
1	GET PART FROM CONVEYOR & WALK TO CRIMPING PRESS	5	---	
				3
2	UNLOAD / LOAD PART IN CRIMPER HIT SWITCH & WALK TO LEAK TESTER	12	10	
				4
3	UNLOAD / LOAD PART IN LEAK TESTER, HIT SWITCH, & WALK TO TEST & LABEL	6	20	
				3
4	UNLOAD / LOAD PART IN TEST & LABEL. HIT SWITCH, & WALK TO PACK STATION.	6	8	
				3
5	PACK PART & RETURN TO STATION 1	14	---	
				5

	Work	Wait	Walk
Total Times:	43	9	18

Figure 6.4   Left side of WCT contains the elements and times.

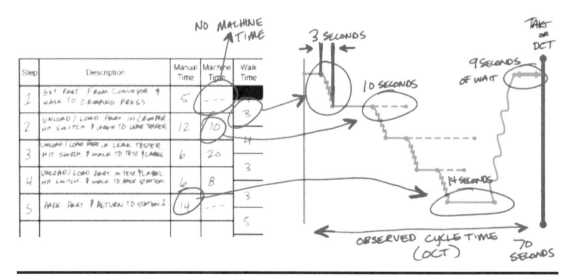

Figure 6.5   Create WCT graph on right side using data from left side.

graph as a red vertical line to denote its importance. In the example in Figure 6.5, the operator cycle is less than the customer TT, which causes a forced wait before the next cycle can begin, and we can see that the system needs improvement. When there is machine time that when drawn would intersect the standard time, the machine cycle dashed line must stop at the red line, and the remainder is *wrapped around* and shown on the leftmost portion of the graph, as shown in Figure 6.6. (The machine time wraps where the cycle begins to repeat.)

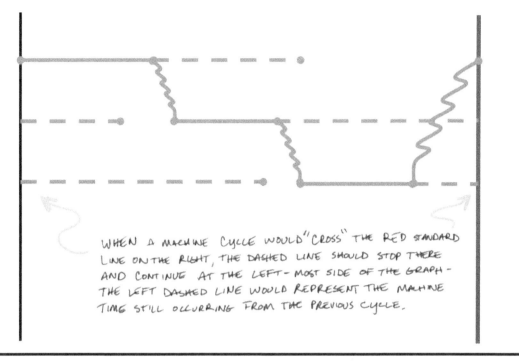

Figure 6.6 Machine times not complete at red line should wrap around.

Another important issue to consider is when there is work that is performed simultaneously while walking. In these cases, the work is not cumulative since it occurs during the time that is already being counted as walking. However, it is still important to graph this correctly so that the condition is not overlooked, and all the work elements are captured. The examples in Figure 6.7 show how to document conditions for work that is less than 1 second as well as 1 second or greater. Note that the time entered in the table is in parentheses to show that it is excluded in the total work time. This condition is also one of the times when work elements become difficult to measure since we have two elements occurring at the same time. In this case, we must exclude the work time and proceed as if it did not occur. Once the other element times are chosen and entered, and the totals (excluding the work while walking time) are calculated at the bottom of the table, then this work time can be entered and enclosed in parentheses.

USE THE COMPLETED WCT TO LOOK FOR OPPORTUNITIES TO IMPROVE

Although a WCT can be drawn freehand on a blank sheet of paper as we have seen, typically, a form is used for making the graph to scale, which helps organize the information for further analysis. Notice that the graph on the

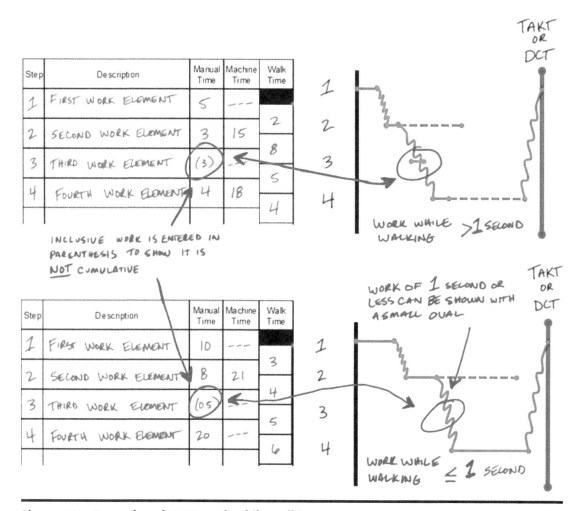

Figure 6.7   Examples of WCT work while walking.

right-hand side is much easier to read, and opportunities for improvement can be quantified much more clearly. Refer to Figure 6.8.

As we have learned, the WCT is a tool for *kaizen*. One source of clues in our search for potential improvement possibilities is in the appearance of the work element graphic. When analyzing a WCT, there are some common questions that often arise. Referring to Figure 6.9, although the following table does not represent all the questions that might be important, we can commonly use them as starting points in our efforts to identify problems and opportunities. Note that the observed cycle time (OCT) refers to a sample worker cycle that was actually observed and measured.

As we consider the starting point questions for our example, we will notice that the double-ended arrow on the right side of the graphic indicates that the worker is forced to wait after returning to step 1 at the end of the process cycle—because a part is not yet available at the conveyor coming from the feeding process. At the bottom of the WCT, it shows that this forced wait is 9 seconds. Even though the worker completed the cycle in 61 seconds, a part is not available at the conveyor pickup point to begin the next cycle until 9 seconds

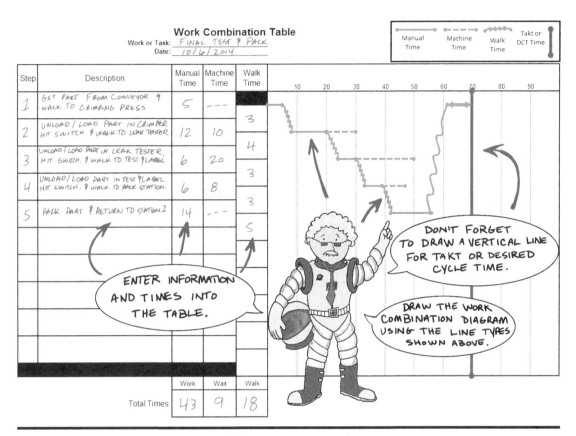

Figure 6.8 Observe, measure, and document accurately.

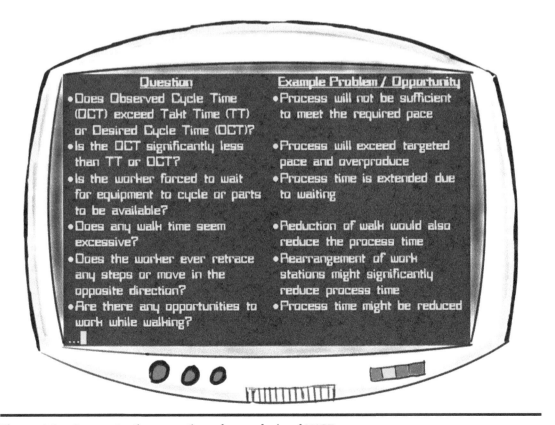

Figure 6.9 Some starting questions for analysis of WCT.

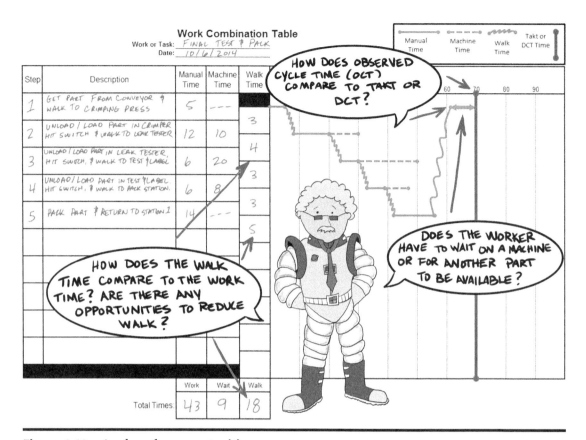

Figure 6.10   Analyze for opportunities.

later forcing the worker to wait and causing the total cycle to be 70 seconds. In this case, 70 seconds happens to be TT, although we can plainly see what appears to be an opportunity to beat this target, as pointed out in Figure 6.10. Accurately portraying the problems on the WCT allows us to search out and quantify such opportunities.

"THINGS ARE OFTEN DIFFERENT IN THE REAL (WORK) WORLD."

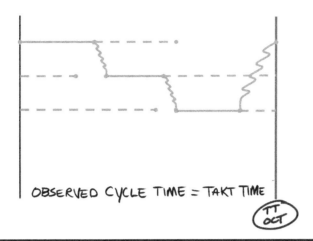

OBSERVED CYCLE TIME = TAKT TIME

Figure 6.11   Example of WCT with OCT = TT.

Meanwhile, Here in the Work (Real) World

It is critical that the WCT illustrate the problems accurately before the analysis is started; otherwise, they cannot be correctly studied. But, before we consider the following examples and the common problems that they reflect, we should be familiar with what a normal WCT graph might look like (see Figure 6.11). Notice that the solid lines denoting manual work somewhat resemble a set of stairs. The downward stair-step pattern indicates that the worker continues in the same direction as when the cycle started. If the steps make an upward pattern, this indicates that the worker is going toward a location that he or she has passed by or has been to previously during that particular cycle—such as returning to a previous station or most notably where the order of process steps does not reflect the geographic order of the workstations. And, we have learned that the machine cycle time *breaks* at the TT or DCT target line and then sort of wraps around from the vertical line on the left side. This represents the machine time from the previous cycle.

Real-World Problem: OCT > TT

When the OCT is greater than the standard time, as shown in Figure 6.12, it is obvious that the worker will not be able to meet the customer requirements if the observed cycle is not just an anomaly. This is one of the most common problems that are encountered in real life. To resolve this problem, it is necessary to increase the time that is available for the work (thereby increasing the TT) or reduce the work element times through *kaizen* (preferred).

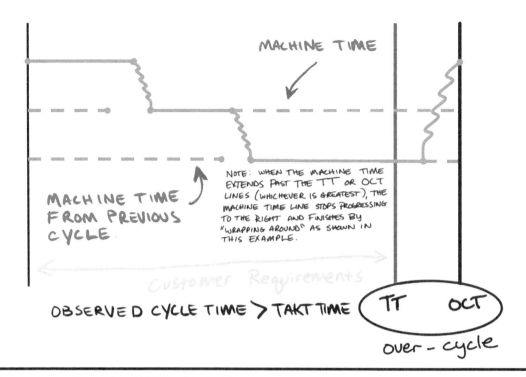

Figure 6.12   Example of WCT with OCT > TT.

"THERE CAN BE A LOT OF DIFFERENCE BETWEEN THE WORLDS."

Real-World Problem: OCT < TT (and Forced Wait at End of Work Cycle)

When the OCT is shorter than the customer requirement (TT), the worker is going to overproduce. Under normal conditions, if the work cycle is shorter than needed, it is necessary to decrease the time that is available for the work (thereby decreasing the TT) or add forced wait at the end of the work cycle to eliminate the opportunity to overproduce. The latter condition may not be sustainable over time and usually results in overproduction or a decrease in the work pace (which is difficult to resolve). However, there are some instances in real life where the worker is forced to wait before beginning their next cycle due to the availability of a part or the completion of a machine cycle. An example of waiting on a part before starting the next cycle is shown in Figure 6.13. In either case, *kaizen* is necessary in order to eliminate the waiting.

Real-World Problem: Forced Wait during Work Cycle

In the previous example, we saw forced wait at the end of the work cycle. Sometimes, forced wait occurs during the work cycle. A common example of this problem happens when a machine cycle from the previous worker cycle has not been completed by the time that the worker arrives back at the workstation of that particular machine. In this case, the worker must pause until the machine cycle is complete. An example is shown in Figure 6.14.

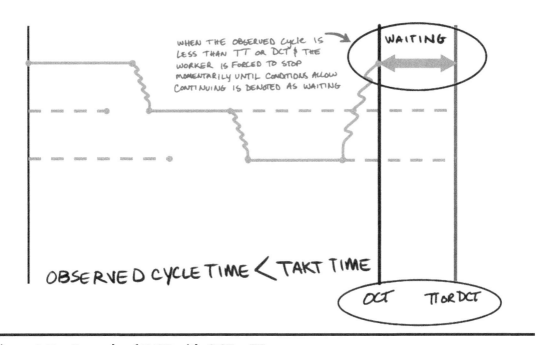

Figure 6.13   Example of WCT with OCT < TT.

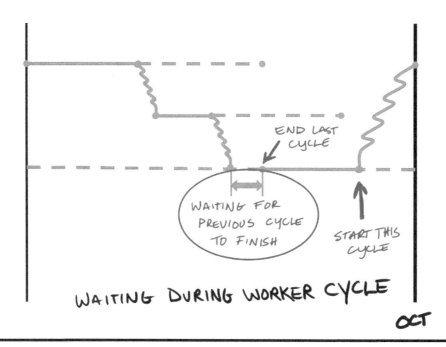

Figure 6.14   Example of WCT with wait during worker cycle.

Real-World Problem: Showing Work Elements with No Walk between Them

In some instances, it may be necessary or desirable to break down the work elements at a location when there is no walking connecting the steps. This is shown as a vertical line, and the steps are numbered as shown to identify that they are part of a single larger work element. Refer to Figure 6.15.

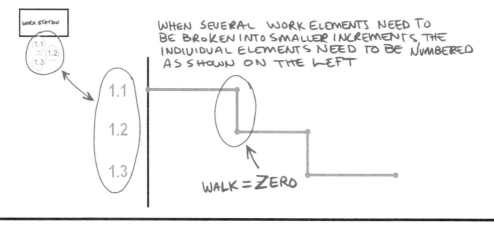

Figure 6.15   Example of WCT with walk = zero.

"SOMETIMES A DOSE OF REALITY IS NEEDED."

A Dose of Reality: Some Problems Are Much More Serious

The real-world issues and problems up to this point have been pretty simple to work around. However, as we know, the real (work) world is not so easy. We do not always have the luxury of being able to do things the ideal way. Often, we are faced with real-world obstacles that may not be economically feasible to overcome—we may not have the money for the ideal equipment, the floor space for the ideal layout, the ideal sequence of process steps between different products/services, or the ideal work/machine cycle time balance for labor optimization. But, if we let that stop us, we may not be competitive in the real world. So, even though the standardized work may not be ideal, it is still very important to accurately and thoroughly document it—and especially the problems—so that we can evaluate future *kaizen* opportunities.

"IN REALITY, SOME PROBLEMS ARE MUCH MORE SERIOUS."

Real-World Problem: Work Sequence Does Not Match Geographic Sequence

When the work sequence and the physical location do not match up, the worker may actually walk back and forth over the same ground during a cycle. If this is not documented properly, the WCT will not illustrate the problem accurately. In Figure 6.16, the example used earlier on how to create a WCT has had the work sequence changed. Now, the operator must go to the grind process before the mill process. If the work sequence was simply graphed on the WCT with no regard for location, the resulting table would have appeared no different from the original example. Notice how the WCT graph now shows the back-and-forth movement of the work sequence.

Real-World Problem: Worker Returns to Same Location during Cycle

Sometimes, a worker must return to a location before the completion of a cycle. In such cases, the different steps should be grouped together as a single geographic zone. This complicates the WCT so that the data table will require multiple entries, as illustrated in Figure 6.17—notice the additional column that is labeled "Zone."

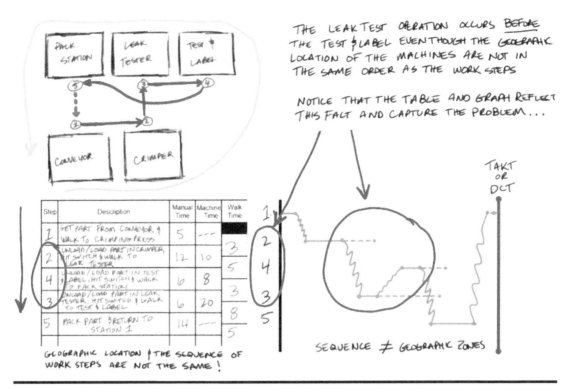

Figure 6.16   Example of sequence ≠ geographic zones.

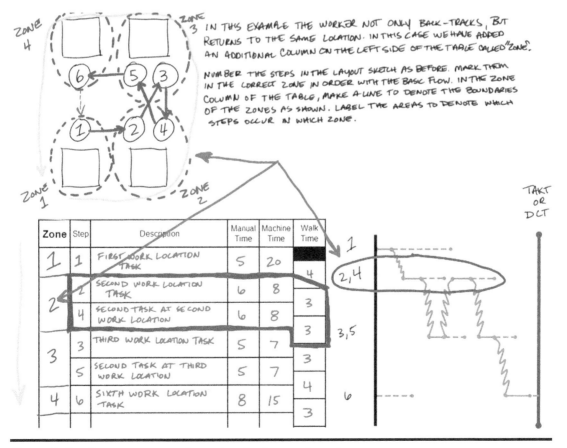

Figure 6.17   Return to a previous step during work cycle.

Real-World Problem: Parallel Machines Due to Excessive Machine Cycle Times #1

Some production processes require multiple machines in parallel due to a large mismatch between the process cycle time and the customer TT. When this occurs, the worker must load a different machine in the parallel set on consecutive passes. In the example shown in Figure 6.18, the second step in the work cycle is composed of two machines in parallel, and the cycle does not repeat until two complete passes are made through the cell. This complicates the WCT and often results in variation between the subsequent parts. However, if the problems are captured correctly, we can better quantify the benefits of the *kaizen* activities in order to help justify the resources that are required.

Some Problems Seem Impossible to Resolve

If we have a solid understanding of the principles upon which Lean and, more specifically, standardized work are based, it can be possible to head off many problems upfront before the system is in place. However, some issues just cannot

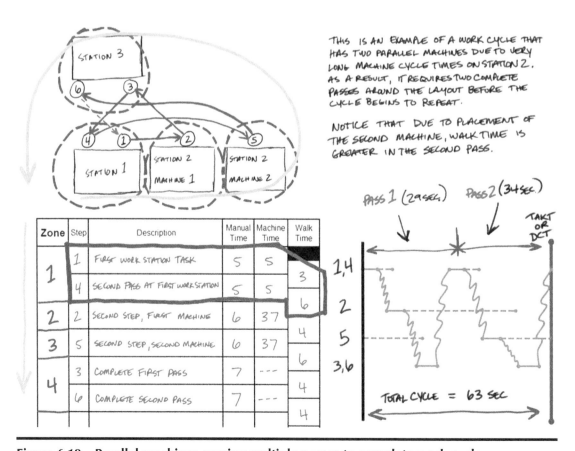

Figure 6.18 Parallel machines causing multiple passes to complete work cycle.

be avoided. Often, the main reason for many of the problems and issues high-lighted in this chapter is the high cost of some process steps—whether based on the high cost of equipment or of specialized skills.

In this next example, in order to try and avoid buying a sixth wire bonder machine, the equipment team sought to utilize five wire bonders, each with dual-part nests. These dual nests allowed for some concurrent processing on one part, while the machine bonded the other part because the sixth machine would not be fully utilized. This worked to some extent, but we leave it up to the readers to decide if the final situation would have been better with six wire bonders with single nests. Whatever your opinion on the wire bond process decision, one thing is certain—it is very important to be able to accurately capture and document the standardized work using a WCT when applicable.

"DON'T LET A DAUNTING PROBLEM DISCOURAGE YOU."

Real-Life Example: Parallel Machines Due to Excessive Machine Cycle Times #2

In the example in the "Parallel Machines Due to Excessive Machine Cycle Times #1" section, it was shown that two parallel machines at the second step complicated the process such that it required two passes to complete a work cycle. In real life, this is an unfortunate possibility in many industries. The following example is derived from an actual application in the electronics industry.

Only one worker's cycle from the cell is shown for simplicity. The wire bond machines on the right side of the figure have a cycle time that is much longer than the customer TT—so much longer in fact that with the overhead time required to slide a nest into position, clamp the part, and then locate the numerous bonding positions with the machine vision system, a sixth machine was even considered. It was finally determined that five machines could suffice if double nests were installed on the machines to reduce overhead time based on the cost of another machine. The dual nest design allowed some concurrent machine clamping and computer vision location processing to be accomplished, while the machine bonded the part in the previously loaded nest. The detail highlights are shown in Figure 6.19.

It is not difficult to determine the number of passes that are required before the work cycle repeats. Five bonders with two nests each results in 10 passes. This means that the worker will load/unload both the oven and the continuity tester 10 times before cycling through all the wire bonder nests. We consider each bonder as a single geographical zone, and, as a result, the completed layout sketch appears, as shown in Figure 6.20.

Even a printed WCT form is difficult to use for this situation due to the complexity that is caused by the parallel machines. Accuracy in capturing and showing the problems is a critical aspect of the WCT. For this real-life example, notice how the data table becomes quite large. In Figure 6.21, we will notice that the left-side portion of the WCT is quite large, and, in order to even have a chance

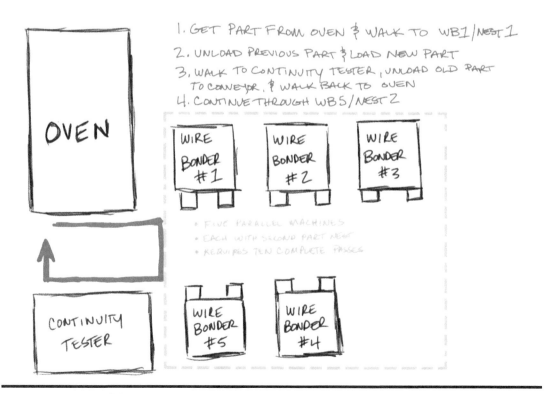

Figure 6.19   Real-life example of multiple parallel machines.

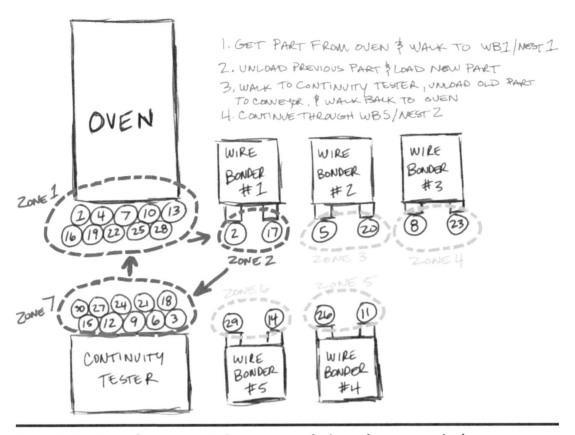

Figure 6.20   Secondary part nests increase complexity and passes required.

of showing the information in this field guide, the left side had to be broken into two pieces and shown side by side. An important point to note with this example as we go forward is that even though the task of capturing all the details accurately seems nearly impossible, it is still critical that we follow through and develop an accurate portrayal of the situation using the WCT. Otherwise, we can miss some very important opportunities to improve.

"DON'T GET HUNG UP IN THE 'RULES'."

Zone	Step	Description	Manual Time	Machine Time	Walk Time
1	1	UNLOAD OVEN - PASS 1	3	---	3
	4	UNLOAD OVEN - PASS 2	3	---	5
	7	UNLOAD OVEN - PASS 3	3	---	7
	10	UNLOAD OVEN - PASS 4	3	---	6
	13	UNLOAD OVEN - PASS 5	3	---	4
	16	UNLOAD OVEN - PASS 6	3	---	3
	19	UNLOAD OVEN - PASS 7	3	---	5
	22	UNLOAD OVEN - PASS 8	3	---	7
	25	UNLOAD OVEN - PASS 9	3	---	6
	28	UNLOAD OVEN - PASS 10	3	---	4
2	2	LOAD WIRE BONDER 1 - NEST 1	4	140	4
	17	LOAD WIRE BONDER 1 - NEST 2	4	140	4
3	5	LOAD WIRE BONDER 2 - NEST 1	4	140	6
	20	LOAD WIRE BONDER 2 - NEST 2	4	140	6
4	8	LOAD WIRE BONDER 3 - NEST 1	4	140	8
	23	LOAD WIRE BONDER 3 - NEST 2	4	140	8

Zone	Step	Description	Manual Time	Machine Time	Walk Time
5	11	LOAD WIRE BONDER 4 - NEST 1	4	140	5
	26	LOAD WIRE BONDER 4 - NEST 2	4	140	5
6	14	LOAD WIRE BONDER 5 - NEST 1	4	140	3
	29	LOAD WIRE BONDER 5 - NEST 2	4	140	3
7	3	UNLOAD/LOAD CONTINUITY TESTER - PASS 1	4	5	3
	6	UNLOAD/LOAD CONTINUITY TESTER - PASS 2	4	5	3
	9	UNLOAD/LOAD CONTINUITY TESTER - PASS 3	4	5	3
	12	UNLOAD/LOAD CONTINUITY TESTER - PASS 4	4	5	3
	15	UNLOAD/LOAD CONTINUITY TESTER - PASS 5	4	5	3
	18	UNLOAD/LOAD CONTINUITY TESTER - PASS 6	4	5	3
	21	UNLOAD/LOAD CONTINUITY TESTER - PASS 7	4	5	3
	24	UNLOAD/LOAD CONTINUITY TESTER - PASS 8	4	5	3
	27	UNLOAD/LOAD CONTINUITY TESTER - PASS 9	4	5	3
	30	UNLOAD/LOAD CONTINUITY TESTER - PASS 10	4	5	3
			Work	Wait	Walk
		Total Times.	110	---	132

Figure 6.21   WCT data table for five parallel wire bonder example is huge.

Do Not Get Hung Up: Some Rules Are More Like Guidelines

The graph portion of this WCT becomes extremely complex and problematic. In this instance, two printed WCT forms had to be cut and taped together to be able to capture the 10-cycle wave form. In addition, there are other problems with capturing a complete cycle. The first is that the wavy-line technique normally used to denote walk becomes very difficult to show without distorting the

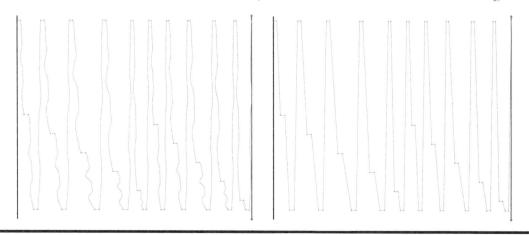

Figure 6.22   Comparison of wavy-versus-straight walk lines on complex example.

view of a complete cycle. In these cases, it is best to use straight lines for the walk. It makes the graph easier to read and allows the use of a straight edge to ensure accuracy. A comparison is shown in Figure 6.22.

"EVEN WHEN THE GOAL SEEMS IN SIGHT WE ARE NOT DONE."

Sometimes, We Outsmart Ourselves

Another problem arises in cases with multiple machine nests, as in this example. Notice that the machine time from the first cycle of a specific wire bonder is not yet complete, and the machine time for the second nest would land on top of the first nest work and machine time lines. This requires that we alter the normal convention of plotting the same geographic region on the same line to plotting the second nest on the next line, as shown in Figure 6.23. The second nest line is plotted in green to highlight this fact. This causes the graph of a complete work cycle to be even more complex.

However, if we are diligent and careful to capture the problems correctly and accurately, the results are worthwhile. Note how the problems can easily be recognized from a visual perspective. But, now, we notice that there is considerable part-to-part variation that is introduced *by design*, as shown in Figure 6.24.

The multiple machines obviously cause the worker to walk different distances, as shown in this very complex real-life example—thus, the individual part cycle varies from 21 to 29 seconds. However, there are other issues that contribute to this variation. Consider that each time the worker arrives at the group of wire bonder machines, he or she must be looking ahead and trying to judge where

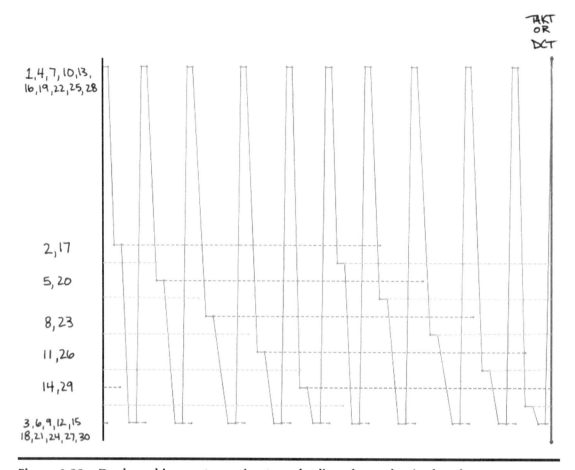

Figure 6.23   Dual machine nests causing two plot lines for each wire bonder.

he or she will be walking to as he or she leaves the oven area. It is possible for the worker to remember where he or she must travel to next by remembering the location of the last pass, but this can get very complicated based on the length and number of interruptions in between passes, which can lead to additional variation and additional worker fatigue.

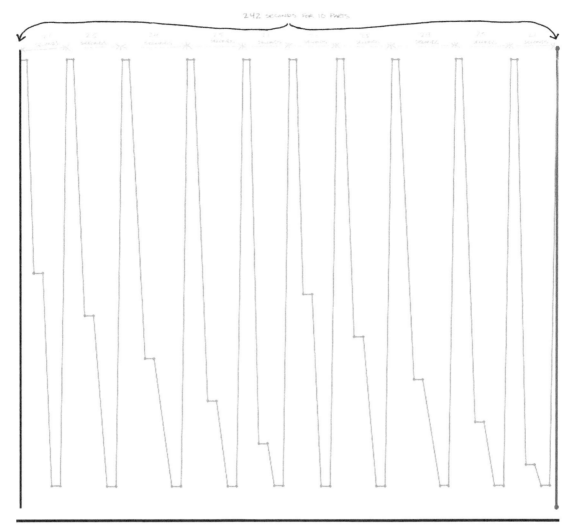

Figure 6.24   Parallel machines with dual nests causing variation by design.

Some Food for Thought

The issues and problems of the previous real-life examples provide us with much food for thought. For example, the dual nest design used was intended to prevent having to purchase a sixth wire bonder. However, by understanding the results of such a decision, it might not have saved that much money in the long run. Consider that the dual nest design makes each wire bonder much more complex and adds extra machine downtime opportunities due to the nests. There are many other issues like this that can be investigated and analyzed. Consider also that, early on, we indicated that this example is only a portion of a larger cell—thus, the variation between subsequent parts in this section of the work flow will directly impact all the other workers. This impact can be significant. The point here is that by understanding the impact that parallel machines and nests can have upon the output of the assembly cell, it becomes possible to have more in-depth discussion with the machine designers and maybe head off some of the issues before the system is built. Significant *kaizen* after the fact might prove to be prohibitively expensive.

In the wire bonder example, we notice that the walk and the load/unload times were much longer than this last example. Although it is outside the scope of this field guide, there are several reasons for this. The work was better designed—counterclockwise-versus-clockwise flow for one-handed loads/unloads and the walk distances that are much less than before, to name a couple of reasons. Yet, we still see part cycle time variation from 37 seconds for one part to 43 seconds for the other. As indicated by the WCT, the singulation process being a part of the cell causes a lot of this variation by design. Even though the average part cycle time is 40 seconds, the worker must keep track of not only the relationship of his or her work flow with the parallel machines but also the status of the singulate and programming processes. The more complicated the worker's flow, the greater the variation and thus the greater chance for error. This example required only four passes around the cell before the cycle repeats. The four parallel machines are one rhythmic variation in the cycle. The two-piece flow portion of the earlier process steps is another. Since one is a multiple of the other, the smaller variation easily *fits* inside the larger variation.

"WASTE IS EVERYWHERE!"

Watch Your Step: Waste Is Everywhere

In the following example, there is an added twist to the issues and problems that were covered so far. Often, there are definite advantages to incorporate more process steps into the work cell or work flow in order to allow more efficient utilization of the workers. However, in the next case, we see that some process steps do not lend themselves to combination very well. For this next example, it is a change in flow that results in skipping a step on alternating passes.

Real-Life Example: Changing the Unit of Flow during the Worker Cycle

Another issue that can really stand out on a properly drawn WCT is a change in the unit of flow. In the following real-life example, a 2-up circuit board array is singulated, and both of the single printed circuit boards (PCBs) are programmed in parallel in a single machine with four nests. Note that when we use the term *2-up*, we mean that there are two identical circuit boards that are printed on a single board with the intention of separating (singulation) them at a later time so that they can process through an automated surface mount equipment as if they were a single circuit board—this is a common practice in order to utilize standard equipment for custom circuit board sizes. The remainder of the cell processes one PCB at a time. There are four test machines in parallel in the latter part of the cell, as shown in the layout sketch in Figure 6.25.

On the first part of the cycle, the worker takes and singulates a 2-up PCB. While still holding a part in each hand, the worker then walks to the programming station. The programming station has four nests; two are empty, and two have parts that have already completed programming. The worker places the

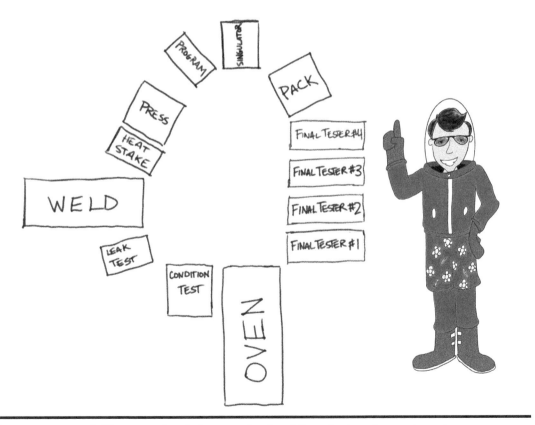

Figure 6.25   Real-life example of change of unit of flow during the work cycle.

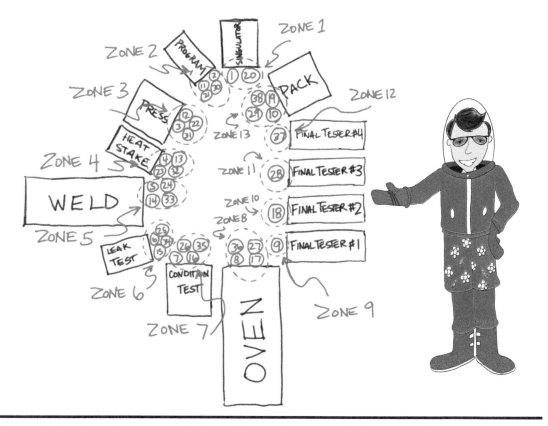

Figure 6.26   Singulate process in the cell complicates parallel machines issue.

two parts into the empty nests. Next, he or she unloads one programmed part and moves to the compliant pin press. The remainder of the pass is completed with a one-piece flow. However, once a part is packed, the worker must skip the singulate process, go directly to the programming station, and take the remaining previously programmed part. (The other two parts are still being programmed.) Again, this pass continues with one-piece flow. After the pack process, the singulate cycle begins to repeat, and the worker again gets a 2-up PCB and proceeds to the singulate process. This results in a very complex sequence numbering of the zones, as shown in Figure 6.26.

Since there are parallel machines at the final test step, the data table portion of the WCT becomes very large, as we learned in the previous real-world example. This often requires that we cut and paste together larger versions of a paper WCT form. If we try to develop the WCT on a computer, it becomes very difficult to ensure that all the steps are captured and recorded in the correct order, especially the walk times. As before, the data table must be cut and pasted to be readable in this field guide. Refer to Figure 6.27.

The designed-in variation begins to take form as we develop the WCT graph. As we can see, the two- to one-piece flow problem can be shown if the WCT is drawn properly. The WCT in Figure 6.28 shows the first two parts.

Zone	Step	Description	Manual Time	Machine Time	Walk Time
1	1	SINGULATE 1st 2-UP PCB ARRAY	2	2	1
	20	SINGULATE 2ND 2-UP PCB ARRAY	2	2	1
2	2	PLACE SINGULATED PCBS IN NESTS 3&4 &TAKE PROGRAMMED PCB FROM NEST 1	3	14	2
	11	TAKE PROGRAMMED PCB FROM NEST 2	1	---	2
	21	PLACE SINGULATED PCBS IN NESTS 1&2 &TAKE PROGRAMMED PCB FROM NEST 3	3	14	2
	30	TAKE PROGRAMMED PCB FROM NEST 4	1	---	2
3	3	UNLOAD/LOAD PRESS & GET/LOAD CASE - PASS 1	3	10	1
	12	UNLOAD/LOAD PRESS & GET/LOAD CASE - PASS 2	3	10	1
	22	UNLOAD/LOAD PRESS & GET/LOAD CASE PASS 3	3	10	1
	31	UNLOAD/LOAD PRESS & GET/LOAD CASE PASS 4	3	10	1
4	4	UNLOAD/LOAD HEAT STAKE - PASS 1	2	10	1
	13	UNLOAD/LOAD HEAT STAKE - PASS 2	2	10	1
	23	UNLOAD/LOAD HEAT STAKE - PASS 3	2	10	1
	32	UNLOAD/LOAD HEAT STAKE - PASS 4	2	10	1
5	5	UNLOAD/LOAD WELDER - PASS 1	2	25	1
	14	UNLOAD/LOAD WELDER - PASS 2	2	25	1
	24	UNLOAD/LOAD WELDER - PASS 3	2	25	1
	33	UNLOAD/LOAD WELDER - PASS 4	2	25	1
6	6	UNLOAD/LOAD LEAK TEST - PASS 1	2	8	2
	15	UNLOAD/LOAD LEAK TEST - PASS 2	2	8	2
	25	UNLOAD/LOAD LEAK TEST - PASS 3	2	8	2
7	34	UNLOAD/LOAD LEAK TEST - PASS 4	2	8	2
	7	UNLOAD/LOAD CONDITION TEST - PASS 1	2	18	2
	16	UNLOAD/LOAD CONDITION TEST - PASS 2	2	18	2
	26	UNLOAD/LOAD CONDITION TEST - PASS 3	2	18	2
	35	UNLOAD/LOAD CONDITION TEST - PASS 4	2	18	2
8	8	UNLOAD/LOAD OVEN - PASS 1	2	---	2
	17	UNLOAD/LOAD OVEN - PASS 2	2	---	2
	27	UNLOAD/LOAD OVEN - PASS 3	2	---	3
	36	UNLOAD/LOAD OVEN - PASS 4	2	---	4
9	9	UNLOAD/LOAD FINAL TESTER #1	2	150	5
10	18	UNLOAD/LOAD FINAL TESTER #2	2	150	4
11	28	UNLOAD/LOAD FINAL TESTER #3	2	150	3
12	37	UNLOAD/LOAD FINAL TESTER #4	2	150	2
13	10	PACK OUT PART - PASS 1	4	---	1
	19	PACK OUT PART - PASS 2	4	---	3
	29	PACK OUT PART - PASS 3	4	---	2
	38	PACK OUT PART - PASS 4	4	---	3

	Work	Wait	Walk
Total Times	88	---	72

Figure 6.27   Data table portion of WCT becomes larger and more complicated.

Wrap-Up

After the WCT is created and the problems are correctly represented, the tool is now ready for analysis. Just as it is necessary to first stabilize the situation when applying standardized work before sustainable improvements can be made, it is necessary that the graphic on the WCT reflect the real problems and issues so that analysis can be done, and the *kaizen* process can proceed. In the preceding example, a first step might be to try and eliminate the variation due to the change of flow. This would eliminate the 3 seconds of variation between successive passes through the cell. The next step would be to try and verify that the

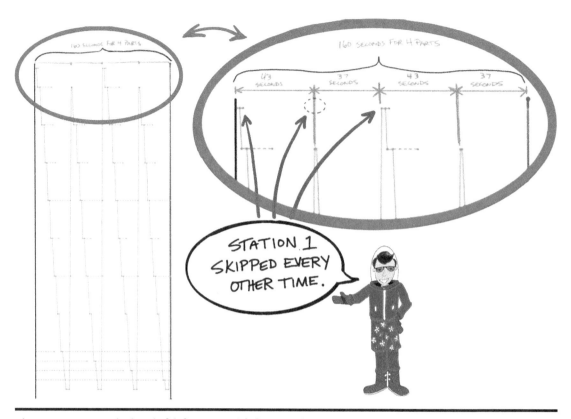

Figure 6.28 Variation is high even with less walk and load/unload time.

time was eliminated if the necessary changes were made. It is very important to find a way to verify that any proposed changes will have the expected results before making any permanent or expensive changes to the system. Otherwise, if the problem is analyzed incorrectly, the situation may have less than expected results or actually get worse.

These last two examples show us that the real world presents us with some very complex challenges. Some people would say that trying to develop an accurate WCT for examples such as these last two is a significant waste of time. We strongly disagree, as we have seen from our examples, that there are significant opportunities for improvement. If we understand how to make and use an accurate WCT, the rewards can be significant as well.

One last point—throughout this field guide, we have tried to show that almost everything can be done by hand with a pencil and paper. However, many people prefer to use computer-based versions of the tools and techniques that were demonstrated. As we have seen in these last two real-world examples, there are many aspects of these methods that are not easily adapted to a computer. This is why a computer-based WCT is insufficient to properly document the issues.

"MOVING ON TO EVEN BIGGER THINGS!"

Chapter 7

Questions for Work Combination Table Review

Introduction

"WE SHOULD EXPECT TO UNCOVER MORE PROBLEMS AS WE GO ALONG."

In Chapter 6, we went through the development of a work combination table (WCT), and we now see that many of the tools are meant to be used in conjunction. This is a very important point to realize as there are quite a few tools, methods, and principles in Lean, so there are going to be many instances where they intersect or otherwise overlap. This is a good thing since Lean is not a science but rather more of a way of thinking. This thinking can be applied to anywhere that there are work processes. Because of this fact, by now, we should have begun to see where the thinking can be expanded to a potentially unlimited number of applications. We say potentially unlimited because we must be careful not to limit the possibilities because of our thinking being too shallow. One need only look at the number of Lean books, conferences, Internet blogs, and other sources to hear about how others have found ways to apply the principles in their own varied applications, so it naturally follows that the tools and techniques would also apply—one just has to think deeply and have an open mind.

Question 1: (Choose one) Which of the following steps is not correct when creating an accurate work combination chart?

1. Start with a layout sketch.
2. Draw a rough diagram based on the layout sketch.
3. Define the read points and capture the times from observed cycles.
4. Enter the past historical standard cycle times into the WCT form.
5. Create the work combination diagram from the data in the table.
6. Use the completed WCT to look from opportunities for improvement.

Use the answer area for question 1 for your answer.

ANSWER AREA FOR QUESTION 1

"SOME PROBLEMS ARE DIFFICULT TO UNDERSTAND AT FIRST."

Question 2: (True or false) The desired cycle time (DCT) or takt time (TT) can be used in the graphical side of the WCT to compare the worker and machine observed cycle times (OCTs).

Use the answer area for question 2 for your answer.

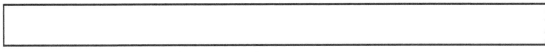

ANSWER AREA FOR QUESTION 2

Question 3: (True or false) When a machine time approaches and extends past the red line, which could be defined as TT or DCT, it is okay to stop this machine element time at the line for the next cycle.

Use the answer area for question 3 for your answer.

ANSWER AREA FOR QUESTION 3

"IT'S NOT ALWAYS CLEAR AT FIRST HOW THE TOOLS CAN HELP US."

Question 4: (True or false) When the OCT is less than the TT or the DCT and the worker is forced to stop momentarily until conditions allow continuing are denoted as waiting.

Use the answer area for question 4 for your answer.

ANSWER AREA FOR QUESTION 4

Question 5a: Complete the work combination diagram using the times recorded in the WCT for question 5a including the work, walk, and wait statistics.

Note: The worker does not have to wait on the wheel test machine cycle for the vehicle assembly. He or she will unload the vehicle assembly from the prior work cycle to be taken to the inspect-and-pack workstation. The worker will then load the new vehicle assembly to the wheel test workstation. At this time, the wheel test process begins, but the worker is able to move on as soon as the part exchange is made in the wheel test part nest.

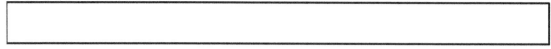

Work Combination Table

Work or Task: Vehicle Assembly
Date: 7/14/2015
Takt Time or DCT: 36 secs.

Step	Description	Manual Time	Machine Time	Walk Time	5	10	15	20	25	30	35	40
1	Body Assembly	4										
				2.5								
2	Wheel Assembly	5										
				2								
3	Final Assembly	6										
				2								
4	Wheel Test	3	3									
				2.5								
5	Inspect & Pack	7										
				3								

	Work	Wait	Walk
Total Times			

WCT FOR QUESTION 5a

The change in the wheel test for this exercise is a common improvement that is made in many applications where variation needs to be reduced. For example, if the worker was expected to wait for a specified amount of time, there would inherently be variation in the wait time since we would be relying on the worker's perception of time, which can vary greatly from person to person. One common improvement from simply asking the worker to pause or wait for a specific amount of time is to add some way to recognize that the correct amount of time has elapsed and is a visual or audible cue such as a light or a bell. This is similar to the cues that we look for when using one of the stopwatch methods. However, even this may not be precise enough in some situations. Another way is to rely on the machine itself since this can be done quite easily with the technology that is currently available. It is important to be able to recognize when there are opportunities for reducing variation that are caused by relying too heavily on the worker for actions that are more suited for a machine. Not only is variation introduced by design when we rely on the perception and reaction time of a worker; it also overburdens him or her with extra work, which can cause unnecessary stress and fatigue.

"SOMETIMES PROBLEMS DON'T SOUND LIKE THAT BIG OF A DEAL!"

Question 5b: Did the OCT for the vehicle assembly work task meet the required TT?

Use the answer area for question 5b for your answer.

```


```

ANSWER AREA FOR QUESTION 5b

The term observed cycle time was introduced to help distinguish a particular cycle time that was actually observed and measured. If we were observing a process and captured the actual time of an actual cycle, we would call this an observation or the OCT. It may or may not be equal to our particular standard time, which would be denoted as TT or the DCT. We would expect our observation samples to vary somewhat as it would be unrealistic to expect otherwise. In our continuous improvement efforts, we strive to reduce the variation and eliminate as many causes of variation that are economically feasible. But, we should also be on the lookout for *special* causes of variation that would not occur very frequently since these observations may often be more economical to ignore and focus on the more common causes of variation in your application.

"DON'T BE AFRAID OF A CHALLENGING PROBLEM!"

Question 5c: (True or false) The machine time for the wheel test step does not have an impact on the ability to meet the required TT.

Use the answer area for question 5c in your answer.

ANSWER AREA FOR QUESTION 5c

It is important to be able to determine what impact a change to the process will have in our continuous improvement efforts. The WCT can help us with this determination. If it accurately reflects the current state, then it becomes much easier to analyze the impact of any changes. However, if it is not accurate or does not exist at all, we can miss many opportunities for improvement. This is because we may not be able to correctly determine the potential positive impact, or, even worse, we may have already given up and assumed that the current state is the best that we can ever achieve. This last problem is more common than we might like to think. In our Lean experiences around the world, we have sadly found this to be true in too many cases. Efforts to produce a *Lean* system often result in people feeling that the culmination of their efforts have resulted in a *win* and then proceed on to the next problem thinking that they are done with this particular problem. If we are diligent and thorough, we can easily avoid this pitfall.

Question 5d: If an improvement opportunity in the layout is recognized and the change shown in the WCT for question 5d could be implemented and verified to eliminate the walk between the final assembly step and the wheel test step, would the ability to meet the required TT be achievable? Redraw the work combination diagram to see.

Work Combination Table

Work or Task: Vehicle Assembly
Date: 7/14/2015
Takt Time or DCT: 36 secs.

Step	Description	Manual Time	Machine Time	Walk Time	5	10	15	20	25	30	35	40
1	Body Assembly	4		2.5								
2	Wheel Assembly	5		2								
3	Final Assembly	6		0								
4	Wheel Test	3	3	2.5								
5	Inspect & Pack	7		3								
		Work	Wait	Walk								
Total Times												

WCT FOR QUESTION 5d

This last exercise is a good example of the power of having an accurate and up-to-date WCT. The left side of the WCT reflects the real-world aspects of the process steps in time and sequence. The right side is intended to provide a visual representation of the left-side data. However, once everything is correct, this does not mean that we are finished. The WCT can now be useful for doing *what-if* scenarios for evaluating the impact of changes to the system that is reflected by the tool. Without this tool, some changes may not be able to be accurately analyzed. If the change can quickly and easily be undone, this *trial-and-error* approach may be fine. However, some changes may not be easily and inexpensively reversed. Having and knowing how to properly use an accurate and up-to-date WCT might prevent us from making a costly mistake.

"DON'T GET DISTRACTED! RESOLVE COMPLEX PROBLEMS ONE STEP AT A TIME!"

Answers for WCT Review

"WE NEED TO RECOGNIZE WHEN TO APPLY TOOLS AND WHEN NOT TO."

Question 1: (Choose one) Which of the following steps is not correct when creating an accurate work combination chart?

1. Start with a layout sketch.
2. Draw a rough diagram based on the layout sketch.
3. Define read points and capture the times from observed cycles.
4. Enter the past historical standard cycle times into the WCT form.
5. Create the work combination diagram from the data in the table.
6. Use the completed WCT to look for opportunities for improvement.

See answer for question 1.

> **The answer is Step 4.** The correct step 4 is to enter the captured data that you observed into the WCT form. It is very important to use observed data from the current conditions and reality of the process that you are analyzing. Historical standard cycle time data is likely to be invalid in the current reality.

ANSWER FOR QUESTION 1

Question 2: (True or false) The DCT or TT can be used in the graphical side of the WCT to compare the worker and machine OCTs.

See answer for question 2.

> **The answer is True.** While Takt Time (TT) is typically used to compare to customer demand, Desired Cycle Time (DCT) can also be used especially if the processes that are being analyzed with a WCT are infrequent processes like setups, change-overs, etc.

ANSWER FOR QUESTION 2

"DON'T BE AFRAID TO TRY A TOOL ON A DIFFICULT PROBLEM!"

There are occasions when the DCT should be the standard because the TT is not useful. A common example where this occurs is when there are multiple products that are run in a work cell, but they have different overall cycle times. As we know, takt time is a standard time that is derived by dividing the number of parts that are required over an allotted amount of time. If we simply total all the volumes of parts that are needed and divide it by the total time that is allotted, the TT standard would have no meaning for each different product being observed. Some might argue that a *weighted* TT value might be derived for each product with a particular overall cycle time, but this can be difficult, time consuming, and very prone to errors—not to mention the fact that the times only apply to a specific mix of volumes. Using the DCT avoids these issues and is why we strongly recommend that the readers have a good understanding of both standards.

Question 3: (True or false) When a machine time approaches and extends past the red line, which could be defined as TT or DCT, it is okay to stop this machine element time at the line for the next cycle.

See answer and additional explanation for question 3.

The answer is False. Refer to **additional explanation for question 3.**

ANSWER FOR QUESTION 3

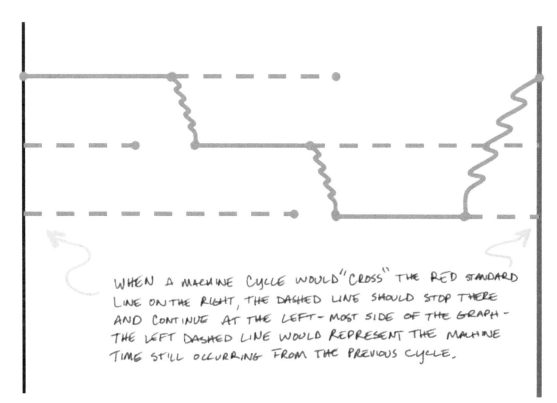

WHEN A MACHINE CYCLE WOULD "CROSS" THE RED STANDARD LINE ON THE RIGHT, THE DASHED LINE SHOULD STOP THERE AND CONTINUE AT THE LEFT-MOST SIDE OF THE GRAPH – THE LEFT DASHED LINE WOULD REPRESENT THE MACHINE TIME STILL OCCURRING FROM THE PREVIOUS CYCLE.

ADDITIONAL EXPLANATION FOR QUESTION 3

The main reason for *wrapping* a machine time once it reaches the vertical standard timeline is that it allows us to determine visually if the machine will be complete by the time the worker returns to that particular machine. It is possible that the machine could still be completing a part from the previous cycle, and this would cause the worker to be forced to wait until the machine completes that part before his or her cycle can continue. But, this may not be a desirable situation. For example, if a worker knows that he or she will be forced to wait somewhere in his or her cycle, it is a common problem for him or her to decrease his or her pace a bit. This type of problem can be extremely difficult to resolve as it can be hard to see if the wait was absorbed by the decrease in pace. Therefore, we should always be on the lookout for instances where the worker must wait.

Question 4: (True or false) When the OCT is less than the TT or the DCT and the worker is forced to stop momentarily until conditions allow continuing are denoted as waiting.

See answer and additional explanation for question 4.

The answer is True. Refer to **additional explanation for question 4.**

ANSWER FOR QUESTION 4

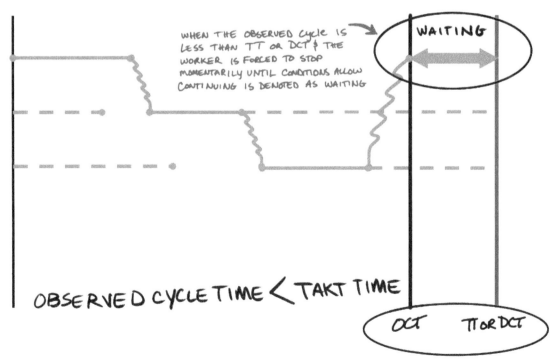

ADDITIONAL EXPLANATION FOR QUESTION 4

Waiting is a common form of waste. Unfortunately, some form of waiting is probably present in just about every process in every industry that we will ever see. Since steps take time (even if it is in very small amounts), this causes a delay before steps later in the process can occur, and, from their perspective, this delay is waiting. However, just because some form of waiting will always exist in our processes, this does not mean that we simply accept it. The concept of continuous improvement, or the *kaizen* attitude, should drive us to constantly look for and implement practical ways to reduce all forms of waste.

As we can see from these last few questions, wait can cause many issues for the worker. One of the main reasons we hear from workers for not wanting to wait is that they feel that it will look like they are not doing their job properly. This is why it is very important to capture the real situation with the standardized work—particularly in the WCT. By capturing the wait properly, the true situation is shown, and the opportunity for improvement is quite clear. If the worker must wait anywhere during his or her cycle, it can then act as a visual cue for our observations. If it is hidden or obscured by a decrease in the worker's pace, the opportunity for improvement might be missed.

"SOMETIMES A TOOL CAN OFFER AN UNEXPECTED SOLUTION TO A PROBLEM!"

Question 5a: Complete the work combination diagram using the times recorded in the WCT for question 5a including the work, walk, and wait statistics.

Note: The worker does not have to wait on the wheel test machine cycle for the vehicle assembly. He or she will unload the vehicle assembly from the prior work cycle to be taken to the inspect-and-pack workstation. He or she will then load the new vehicle assembly to the wheel test workstation. At this time, the wheel test process begins, but the worker is able to move on as soon as the part exchange is made in the wheel test part nest.

See answer for question 5a.

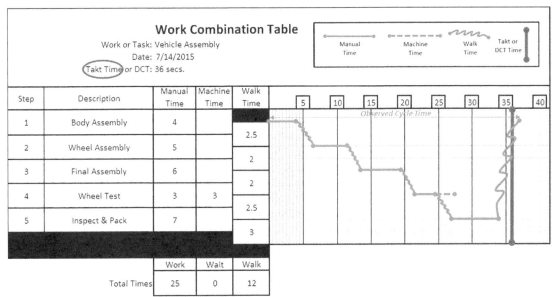

ANSWER FOR QUESTION 5a

"IS THERE MORE TO DISCOVER?"

Question 5b: Did the OCT for the vehicle assembly work task meet the required TT?

See answer and additional explanation for question 5b.

The Answer is No. The WCT shows that the total Observed Cycle Time is equal to 37 seconds as compared to the required Takt Time of 36 seconds. Refer to **additional explanation for question 5b.**

ANSWER FOR QUESTION 5b

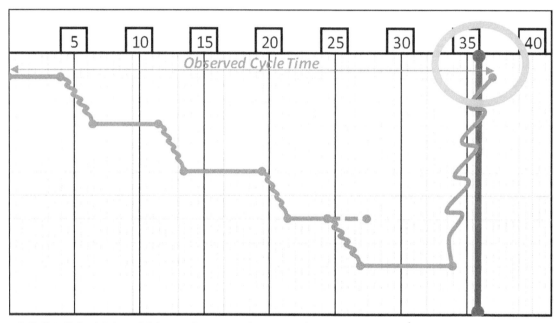

ADDITIONAL EXPLANATION FOR QUESTION 5b

Should we be extremely worried about whether or not we are going to have problems meeting our production schedule if we have a situation such as the one that was presented in question 5b? Not really, because a 1-second difference between the DCT and the TT can usually be resolved quite easily if we are diligent in our continuous improvement efforts. This is where the concept of *kaizen* can really be seen because it is these small increments of improvement that can sometimes help us achieve our goals.

Question 5c: (True or false) The machine time for the wheel test step does not have an impact on the ability to meet the required TT.

See answer and additional explanation for question 5c.

The answer is True. Based on when the worker returns to the Wheel Test workstation, the Machine Time should be finished within plenty of time. This will allow the worker to unload the tested vehicle assembly and load an new vehicle assembly to the Wheel Tester. Refer to **additional explanation for question 5c.**

ANSWER FOR QUESTION 5c

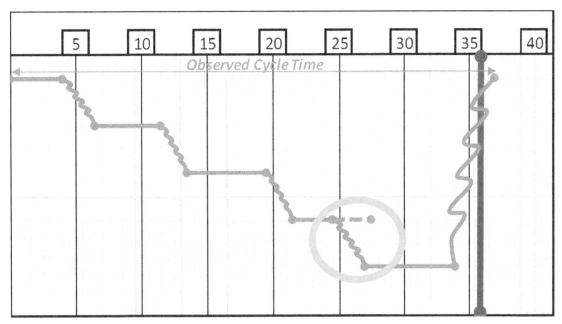

ADDITIONAL EXPLANATION FOR QUESTION 5c

"SOME PROBLEMS ARE ANTICIPATED."

Question 5d: If an improvement opportunity in the layout is recognized and the change shown in the WCT for question 5d could be implemented and verified to eliminate the walk between the final assembly step and the wheel test step, would the ability to meet the required TT be achievable? Redraw the work combination diagram to see.

See answer and additional explanation for question 5d.

The answer is Yes. By consolidating the Final Assembly and Wheel Test Steps into one work area and reducing the walk between the two stations, the new estimated time would be 35 seconds which meets the required Takt Time of 36 seconds. This improvement would actually require the worker to have one second of wait time before starting the next cycle. Refer to **additional explanation for question 5d.**

ANSWER FOR QUESTION 5d

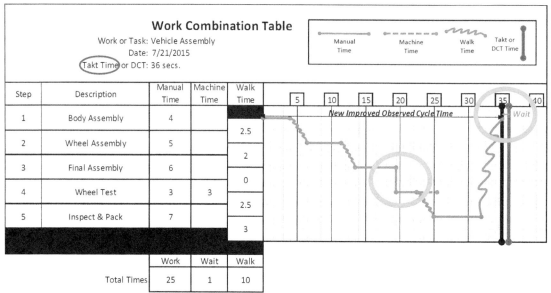

Work Combination Table

Work or Task: Vehicle Assembly
Date: 7/21/2015
Takt Time or DCT: 36 secs.

Step	Description	Manual Time	Machine Time	Walk Time
1	Body Assembly	4		
				2.5
2	Wheel Assembly	5		
				2
3	Final Assembly	6		
				0
4	Wheel Test	3	3	
				2.5
5	Inspect & Pack	7		
				3

	Work	Wait	Walk
Total Times	25	1	10

ADDITIONAL EXPLANATION FOR QUESTION 5d

"SOME PROBLEMS GO AWAY ON THEIR OWN."

As discussed, after question 4, we should always be concerned if there are situations where the worker is required to wait. We know that it can cause other problems as well as additional issues for the worker. Sometimes, it is quite easy for him or her to wait because he or she cannot continue until something else is completed. However, there are also instances where this may not be the case. Consider the situation where the DCT is less than the TT, and the worker is not constrained by having to wait for something to finish before he or she continues. If he or she was able to continue, he or she is overproducing, and this can cause even more problems. Therefore, it is always important to proceed with caution if there is any wait that is built into the worker's cycle.

"LOOKING FOR MORE ADVENTURES!"

Chapter 8

Where Do We Go from Here?

Introduction

"WHERE CAN WE GO FROM HERE?"

In Chapters 1 through 7, we learned how to use the more common standardized work tools and techniques. During our travels, we also learned how to apply these things in many not-so-common situations. These difficult situations came about for a lot of different reasons: some industry specific and others due to design issues. Although you many never experience all of them in your own adventures, you will most likely run across some of them, and it is our hope that by sharing a few of our experiences with various types of problems encountered and how we learned to deal with them, it might prove helpful to others. Unfortunately, it is not a perfect world out there, and the issues and problems we run across are often never mentioned in many published works that we have read over the years. Even some of our friends and teachers from Toyota tell us that they had never seen some of the problems that we discussed in this field guide. This forced us to step back and think deeply about the problems in order to try and find a way to adapt the tools and techniques for a particular issue.

Our intent in sharing some of the difficulties we have encountered is twofold. First, it is not to rob the readers of the learning experience of finding a way of dealing with their issues for themselves but rather to show a few selected examples of our own learning experiences in the hopes that it might give them the confidence to follow through and find a way to adequately deal with the problem or issue. Too many times, we have seen people get discouraged that their particular application does not lend itself easily to using the tools and techniques in this field guide and often give up entirely or at least only utilize a very small percentage of them. As a result, they can miss some great opportunities for improvement. The second intent is to get the readers to start thinking *outside the box* as they say and recognize that these tools and techniques can work in other applications aside from simple work cell applications. But, before we end our adventures together, there are a few other tools that should be mentioned.

"THE SHAPE OF THINGS TO COME."

Up until this point, we have looked for opportunities that focus on reducing waste and variation. Many of the Lean tools and principles were originally developed with the focus of becoming more efficient in order to use the least amount of resources to achieve the customer requirements—a particular number of parts in a particular amount of time, for example. This is because the concept of Lean is derived from the Toyota Production System whose intent is to deliver the best value to the customer in the most efficient manner. In the automotive industry, there is a relatively limited annual demand that is met by a small number of sources—many of which constantly try to gain some market share that eventually must come from the other suppliers. The supply and the demand are not that far apart due to the high cost of manufacturing automobiles. Just because a manufacturer can produce more vehicles, it does not automatically follow that they will sell the additional cars.

However, in some industries, the demand can far exceed the available supply, so if a business is able to produce more without adding too much cost, it is possible to make more profit. During our journeys, it has probably occurred to you that we can also make improvements that are focused on other things such as capacity or output, especially since machines are often part of our process times. In the automobile example, the focus was on increasing profit by producing at the lowest cost—this was because it is very expensive to significantly increase output and sales together. In other words, just because we could make more, it does not mean that we can sell more. We would also need for sales demand to increase accordingly as we have to attract customers from the other suppliers due to the limited demand. Although increased marketing can help increase the demand, it is an expense and not a physical asset. In order to maintain the demand, it may require constant marketing, which increases expenses.

"WHAT IF THE SITUATION COULD BE CHANGED?"

Man–Machine Utilization Graph

But, what if the situation was different? What if increasing the sales demand was not so expensive or already existed? If a business could produce more, it could most likely sell more. In such cases, increasing the output would also be desirable in our search for opportunities to improve. Of course, reducing waste in the work components that result in a shorter overall cycle time would increase the capacity of the system until the practical limit of the system was reached—at that point, additional resources would be required to increase the capacity. But, how do we determine the practical limit of a system? This varies according to the situation. Most systems are constrained by labor or process capacity. In cases such as a work cell where people and machines both impact the overall cycle time, it is not too difficult to determine the practical limits of the system of various conditions. A tool that is very useful in such cases is called a man–machine utilization graph (MUG).

In Figure 8.1, we see an MUG based on the example starting with Figure 6.4. The left side of the MUG is where we total up all the work components for each worker. In the example, there were 43 seconds of total work, 18 seconds of total walk, and 9 seconds of wait. When these are added together, we get 70 seconds, which for this example is the desired cycle time (DCT). If there were other workers involved with the particular system, then they would also be shown.

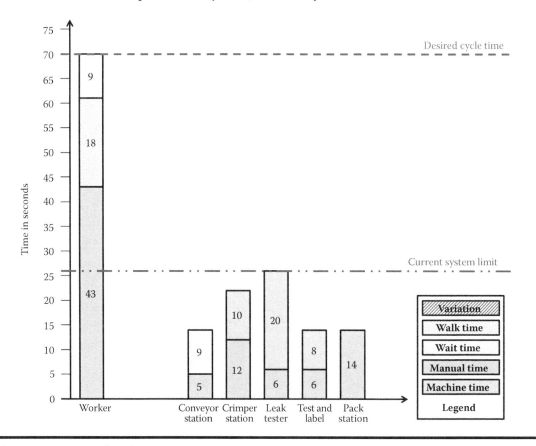

Figure 8.1 Example of an MUG.

The right side of the MUG is where we total up the process times. If we have forced wait, it is also shown if it affects a particular process time. In this example, the wait time is added to the work time for the conveyor station since the worker must wait 9 seconds for the next part to arrive on the conveyor before he or she can begin the next cycle. On the crimper station, we see that there is 12 seconds of work followed by 10 seconds of machine time before the next cycle of this process can begin. For the leak tester, we see that it has 6 seconds of work before the test starts, and then there is 20 seconds of test time before this process can start again. Similarly, the test and label process also has 6 seconds of work time before the 8 seconds of machine time can start. And, finally, we see that the pack station has 14 seconds of manual work before it could be repeated. The process with the greatest total is the current system limit.

"THERE ARE ENDLESS POSSIBILITIES TO IMPROVE!"

The MUG can help understand the capacity as the system currently stands as well as the maximum capacity. It does this by combining two graphs together in order to compare the two main aspects of a system that uses both worker time and machine time. The *man* side is on the left, and the *machine* side is on the right. In Figure 8.1, we have seen that the DCT is limited currently by the man side at 70 seconds. We say that it is limited because it is the greatest total of the two sides. If we were to decrease any of the work component totals in the stack on the left side, we would see a direct decrease in the DCT as well. This helps us look at opportunities to improve that are based on the work components that affect the worker's job. However, the improvements on the left side, even if feasible, are not unlimited. If the total continues to be reduced, eventually, there is a practical limit of how low the stack on the left can go down. This is the point where the stack on the right comes into play—the current system limit.

"THE SKY IS THE LIMIT!"

The current system limit represents the point where under the current conditions, if the man-side components could somehow be reduced sufficiently, one or more machine process times become the limiting factor. Normally, there is only one *highest* stack on the right side constraining the current system limit, but it is possible to have more than one process and labor combination that are tied for the highest stack. In any event, the current system limit would represent the maximum capacity of the system under the current conditions. Notice that when we refer to current conditions, we mean the conditions that this particular graph was created under. If any of the conditions change, the graph must be updated to reflect the new conditions. For example, notice in the example MUG that the worker time is included not only on the man side but also in some of the stacks on the machine side. This means that if the worker time is decreased, not only will the man-side stack change, but, almost certainly, there will be changes on the machine side as well because the current conditions have changed. These changes may or may not reduce the highest stack on the machine side, but the MUG should always be updated if the conditions change. If you are using the MUG for evaluating a possible change, the same situation applies—the difference may be that we must think deeper to understand the effect that a possible change would have to do on each side to see if we think that it is feasible.

"PROBLEMS CAN COME BACK IF WE ARE NOT PAYING ATTENTION!"

Task Summary Sheet

Up until now, we have been discussing tools and techniques that seem to revolve around work cells. However, work cells are not the only place where these tools and techniques can work. They can be adapted to other applications that have similarities with the work cell issues. For example, in Chapter 6, it was briefly mentioned that the work combination table can also be used for situations such as changeovers because the tool helps visualize the effect of each work component on the total end-to-end time of a repeatable work sequence. It is not our intention to try and list all the possibilities out there for these tools and techniques because they are endless. Our intent has been to try and encourage the readers to not only understand how to use and adapt the tools and techniques, but also, by thinking deeper about them, we can find ways to use them for other situations where standardized work and continuous improvement concepts can be applied.

In our experiences in applying standardized work across many work applications, both inside and outside of manufacturing, we have found a definite need for a tool that allows for more details to be made available without becoming too cumbersome. A tool that we have found to be very helpful for standardized work applications is one that we refer to as a task summary sheet (TSS). The TSS is adapted from the Training Within Industry tool that is called the job instruction. An example of a TSS is shown in Figure 8.2.

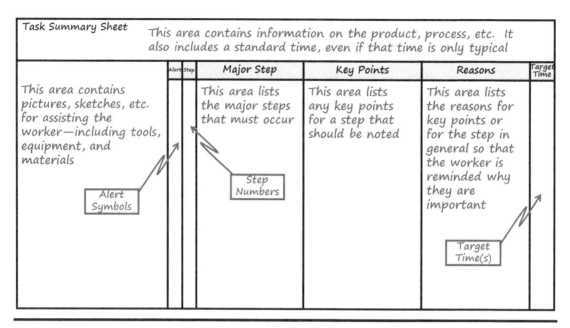

Figure 8.2 Simple example of a TSS.

The TSS is a simple and versatile tool for standardized work applications that do not lend themselves to the tools that we have discussed up to this point. For example, if there were some additional information that the worker needed for the pack station step in the example in Figure 8.2, the tools up to this point have not offered an opportunity to capture and supply this information for the worker. The TSS could be used to supply this information for the worker when needed. A common example is for training purposes. Another example is as a reference for a worker who has not performed the task in a long time and needs to refamiliarize himself or herself with the important details of the task—a good real-life example of this that we have seen is for an annual inventory in a hospital gift shop. The inventory is only performed once per year, and it is difficult to remember all the steps and important issues when a great deal of time passes between cycles. In this real-life example, there are several interim steps that need to be performed, including the setup and preparation of software, file transfers, file renaming, and so on. Without standardized work, this task could not only be very difficult and prone to errors but also a great source of variation. The TSS allows the right information to be available and in the right order.

When used in the context of standardized work, the TSS must also contain a standard time, even if it is only a typical time for reference. In our example in Figure 8.3, if there is no time standard, how long should the task take to complete? How does the worker know if he or she is taking way too long or if he or she is completing the task way too quickly? The time standard serves as a reference point for the worker or observer so that the work under observation can be compared to the standard, just like a cyclic task that is repeated every 30 seconds in a work cell.

Gift Shop Annual Inventory Standardized Work   **Step 1 of 3**

1) Load Scanner with latest UPC codes

Target Time	15 minutes (to get the codes loaded)	

Department:	Gift Shop
Task Name:	Prepare scanner for inventory
Last Update:	January 8, 2013
Updated by:	Tim Martin

	Alert Symbol	Step	Major Steps	Key Points	Reasons	Target Time
List of materials and equipment, including serial numbers if required		1	Prepare data file from "BackOffice" software to be downloaded into the handheld scanner	Make sure that all the newest UPC codes have been entered into the system. To do this, open the BackOffice program. Then follow these steps: - Price Book - POS Utility - Export - Export Handheld Validation File - Item & Price Validation (see loading bar) - Close	If the latest UPC codes are not in the scanner, an error will occur everytime an item is scanned that is not in the system. The steps listed on the left will put the latest UPC codes and prices in a file called Download.txt in the directory C:/1 Handheld Data	
Photos and illustrations of barcode scanner		2	Connect scanner to computer	Make sure scanner is charged and seated in the cradle. Cradle should be connected to AC power through the power supply and connected to the computer USB port with the cable. Ensure that the USB cable is connected to the computer that has the BackOffice software running on it.	Handheld must be properly connected to the computer in order to transfer files between the two devices.	
	◇	3	Configure scanner to receive validation file	Turn the handheld scanner on by pressing and holding the small RED button. Hit [ESC] until the menu options shown are [1. Collect Data] and [2. Utilities]. Use [▲▼] as needed to move cursor to <u>blank line</u> below option 2 and press [←]. Select [1.Transfer Files] and press [←]. Next select [3. Get Lookup] and press [←]. Screen will display [Waiting...Press ESC to Cancel]. Place handheld back into cradle.	Validation file must be downloaded to handheld to check if scanned items are in the database.	
		4	Run "Dlookup.exe" on computer	Double-click on the shortcut icon called "send validation file to hh". Click "OK" in the dialog box to begin send the data file to the handheld scanner. (The transfer will take several minutes).	This file tells the computer to start receiving the data from the barcode scanner.	
		5	When transfer is complete, close dialog boxes on computer	Click on "Cancel" or the "X" to close the dialog box.		
		6	Remove scanner from cradle and configure handheld for inventory scanning	Hit [ESC] as required until the main screen displays [1. Collect Data] and [2. Utilities]. Select [1. Collect Data] and press [←]. Next select [Qty Count L] and press [←]	The [Qty Count L] contains all the UPC codes and prices in the database and any items not in the system can be identified as they are scanned.	15 minutes to this point if no problems
		7	Proceed with inventory process	Face handheld towards UPC barcode and press the large ORANGE button to scan. If the code is in the database, the price will be shown and the cursor located ready to enter the quantity. Enter the quantity and press [←]. (NOTE: If the ORANGE button is pressed again before the quantity is entered the item will be saved with a quantity of zero and <u>will need to be scanned again</u>). If the UPC code is not in the database the error message [Not Found Hit Any Key] will display. Continue until complete.	The UPC code will be read and entered into the data file without human typing error. (When inventory scanning is complete refer to Step 2 of 3: "Prepare upload of inventory to computer" standardized work file).	This time varies based on the amount of inventory to be scanned

Figure 8.3   Example TSS for hospital gift shop annual inventory.

Notice that there are five columns on the right side of the TSS. The first one is for "Worker Alert Symbols," which will be described in more detail in a moment. The next column is for the "Major Step Numbers." The third one is called the "Major Step" and should be a simple text that is used to describe the name of each particular step so that it breaks down the major steps of a process into separate sequential parts. The fourth column is called "Key Points." It is used to list or describe any key points or important issues for that particular step. The last column is simply called "Reasons." It is used when there is a need to explain the reasons why something is done in a particular way. Often, this is used when there is more than one way to do something on this step, and a particular way is necessary. It could be because there is some information that is not obvious to the worker or other observers, or it could be as simple as the workers themselves decided on an arbitrary way just to reduce variation. There may not be a need in all cases to use the reasons column, but most processes have at least some key points or important issues during the steps of the process.

"SOME PROBLEMS WILL COMPETE FOR YOUR TIME."

A very important feature of the TSS is the column for symbols. This allows us to predefine a set of simple symbols that can be added for each step that serve as a visual reminder of important considerations. For example, there may be issues on specific steps for safety, government regulations, local or federal laws, the extra accuracy that is required, and so on that it is important to remind the worker, trainer, auditor, or other observers about. These symbols can be customized as required as long as a legend key is available for reference. Figure 8.4 shows an example of a TSS symbol legend key from a hospital.

◇	Quality check
✚	Safety issue for the worker
●	Standard in-process stock
○	Hand-off point
▽	Critical customer requirement
☺	Patient satisfaction issue
♡	Safety issue for the patient
🔒	Security issue
!	Hospital rule issue
S	Sterile environment issue

Figure 8.4   Example TSS symbol legend key.

The left side of the TSS is intended for pictures, illustrations, tables, etc., that serve to assist the worker in some way. This could include setup, adjustment, tips for the worker, or other helpful information of a visual nature. Sometimes, a picture can help more than several pages of text, and the TSS is not meant to be an extremely detailed tool. Rather, it is intended to be as simple as possible and still convey the important information that is necessary for the worker to accomplish the standardized work.

Normally, standardized work is kept very simple and often only constructed from paper forms with a pencil. However, in real life, there are many applications for standardized work that do not fit the simple application of a cyclic work cell. This does not mean that standardized work is not applicable. The tools that have been introduced throughout this field guide are meant to make the worker's job easier and less variable without adding undue complexity—either in the development or maintenance of standardized work. With the access and availability of computers, printers, wireless networks, digital cameras, tablets, smart phones, and other handheld electronic devices, it is often easier to use electronic files than paper ones. The key is to always remember that the information is for the worker's benefit and that, in order to be able to continuously improve, we may need to update the files often. The updating process for standardized work tools and documentation should not be in any way a detriment to keeping the information accurate and up to date. Otherwise, this will only serve to encourage waiting for significant changes where it is *worth* the time and trouble that it takes to make the updates. This is a major problem anytime the standardized work is not just pencil and paper that can be updated at the *gemba on the fly*.

"JUST WHEN YOU THINK YOU'RE DONE, YOU MUST GO BACK TO THE BEGINNING!"

Take the gift shop example that was described in Figure 8.3. There was an enormous reduction in the man hours that are required for the annual inventory of all the gift shops. The number decreased from 312 hours for the previous year to only 36.5 hours—a decrease of over 88%. Some people would agree that this reduction was so drastic that there would be little reason to spend any more time trying to reduce the time further in the coming years. However, one issue that became very clear, performing the annual inventory with the new method utilizing the barcode scanner, was very different from before, and the interfacing with the computer was not intuitive to the workers, many of whom were volunteers and may or may not be present in the next year's inventory activities. Therefore, the implementation of standardized work was necessary not just to reduce variation from year to year but also to offer a means to retain the information since the process was only performed once per year. By using the TSS as the standardized work form for this task, since the initial change, the time to perform the inventory has further decreased from 36.5 to 13 hours—a decrease of over 64%. In addition, since, by the very nature of standardized work, there must be a standard time for comparison, the workers who might be less familiar with the process can use the standard to help them determine if they are on the right track—similar to the concept of takt that is described in *New Horizons in Standardized Work* (Martin and Bell 2011, pp. 60–64). However, the long time between uses should not be a deterrent to working on improvements. If it is, we miss the opportunity for additional improvement. In a situation like the annual inventory, it is critical to take every chance to improve since it is a long time between opportunities to observe.

Throughout this field guide, we have repeatedly stressed the importance of continuous improvement. We have learned not only how to apply and use several standardized work tools and techniques for reducing variation but also how to use them for making continuous improvements. By now, we should have begun to realize that in order to actually apply the concept of continuous improvement, we must be prepared for continuous changes to our documentation and training. This brings up two extremely important issues: (1) we must be diligent in making sure that we capture every change to the standardized work and update all documentation, worker aids, and training materials that are affected, and (2) we must be timely and thorough in making sure that all changes are communicated to everyone who is involved. Otherwise, the result can be the same as not having any standardized work at all.

"NOW WE MUST PUT WHAT WE HAVE LEARNED TO GOOD USE!"

Chapter 9

Questions for Miscellaneous Tools Review

Introduction

"APPROACHING THE END OF OUR JOURNEY TOGETHER."

The man–machine utilization graph (MUG) is a very useful tool. Just as the work combination table (WCT) can help us identify and analyze improvement opportunities that are related to the individual work components, the MUG can help us identify and analyze opportunities for an entire system or work cell. It allows us to better understand how both the workers and the individual process steps impact one another as well as the entire system. It can quickly show the current limits of the system and help us to intuitively identify opportunities and limitations. In addition, the MUG is much less complex than the WCT and therefore easier, faster, and less prone to errors. Sadly, like the WCT, it is also often overlooked by many, which can mean that there are a lot of missed opportunities for improvement. However, for those who are diligent and willing to put the extra effort into developing WCTs and a MUG for their system, the rewards can be well worth the effort.

Question 1: From the WCT for question 1, develop the man–machine utilization graph (MUG for question 1).

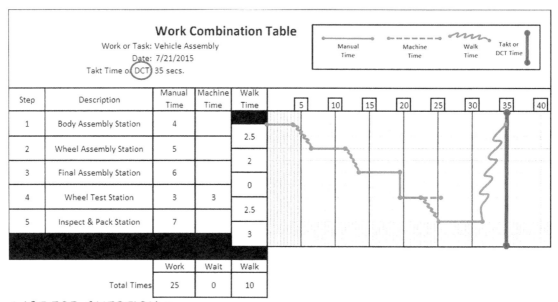

WCT FOR QUESTION 1

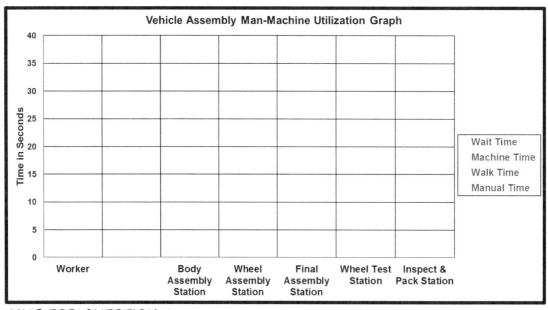

MUG FOR QUESTION 1

"LET'S CONSIDER SOME OF THE THINGS WE HAVE LEARNED."

Question 2: What is the maximum capacity per hour of the assembly system in the completed MUG for question 1?

Use the answer area for question 2 for your answer.

ANSWER AREA FOR QUESTION 2

This example is for a single worker in a work cell. There are many systems that utilize multiple workers, and the method is the same. The only difference is that there will be more workers, which is shown on the left side of the graph. The worker interaction with each machine or process step time is completed just as it is for one worker. Although, for simplicity, we show only the work, walk, wait, and machine times, it is okay to include other times if they also impact the system such as variation. Another example is delays that are brought about by external inter-lock times such as those that are caused by interrupting a machine cycle during the work cycle by a material replenishment worker or an automated guided vehicle (AGV). If the time impacts the worker and/or the process time, it is a candidate for inclusion on the MUG.

Question 3a: If we had the ability to staff a worker at each station in the assembly system with only adding 2 seconds to each workstation for part handling between stations, complete the MUG for question 3a.

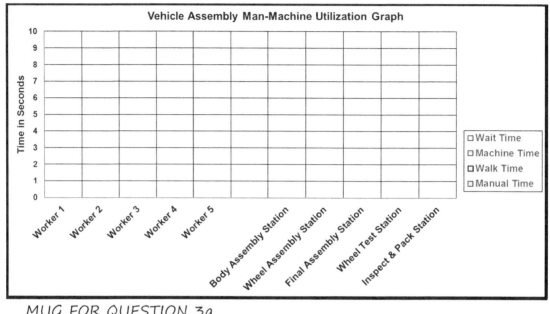

MUG FOR QUESTION 3a

Question 3b: What would be the maximum capacity per hour from the completed MUG for question 3a?

Use the answer area for question 3b for your answer.

ANSWER AREA FOR QUESTION 3b

"WHAT DID WE DISCOVER?"

Question 4: Why is the walk time not shown on the right side of the man–machine utilization chart?

Use the answer area for question 4 for your answer.

ANSWER AREA FOR QUESTION 4

Walking is a big part of a lot of processes. Although it adds no value, it is often a necessary part of standardized work. It is very important that we have a good understanding of the effects that it has and how it is used in the various standardized work tools. Since things are never perfect, trying to reduce one type of waste often results in a compromise with another form of waste. Walk is a good example of this. We often have to add walking in order to not waste another worker's time. However, we are not saying that walking is unavoidable, only that we often must be keenly aware that we may have to make some trade-offs on our journey to improve the overall system.

"SOME THINGS ARE VERY FAMILIAR."

Question 5: (Choose one) Which of the following items below should not be included on a task summary sheet (TSS)?

1. Pictures and sketches.
2. Key symbols.
3. Step numbers.
4. Major step descriptions.
5. Key points.
6. Reasons.
7. Time.
8. All of the above items should be included on a TSS.

Use the answer area for question 5 for your answer.

ANSWER AREA FOR QUESTION 5

"REMEMBER TO THINK DEEPLY ABOUT PROBLEMS."

Answers for Miscellaneous Tools Questions

"WE HAVE TO LEARN TO RECOGNIZE OPPORTUNITIES."

The tools and techniques that have been introduced in this field guide are not all that unique. Many have been around a long time and may be known by many other names. The important thing is to understand how and when they can be applied to help us in our efforts of continuous improvement. By now, it should become apparent that the application of these tools and techniques is much wider than just simply work cells in a manufacturing setting. They can be used in some manner in basically any situation where work is involved. Once we recognize an opportunity, the next issue is finding a way to adapt it if necessary. It is beyond the scope of this book to try and list all the possible applications as they are most likely endless anyway. One of our goals was to try and get the readers familiar with how these tools and techniques are used in the simple manufacturing applications so that they could grasp the ideas behind them and thus provide a solid foundation for adapting them to whatever application that was required. They may not look exactly like the ones in this field guide after they are adapted, but that does not matter as long as they help the readers in their own journeys to continually improve.

Question 1: From the WCT for question 1, develop the man–machine utilization graph (MUG for question 1).

See completed MUG for question 1.

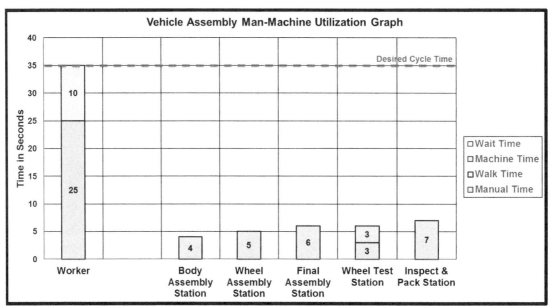

COMPLETED MUG FOR QUESTION 1

Notice how easy it is to develop the MUG if we have a good and legible WCT to work with. It is possible to produce an MUG without one, but if we have an accurate and up-to-date version, the development of the MUG is very quick and simple.

Question 2: What is the maximum capacity per hour of the assembly system in the completed MUG for question 1?

See answer for question 2.

> As shown in the Man-Machine Utilization Chart, the worker is the resource with the maximum cycle time for one complete part work cycle. This time is 35 seconds. So, 3600 seconds per hour divided by 35 seconds per one part work cycle = **102.85 parts per hour.**

ANSWER FOR QUESTION 2

Once we answer question 2 above using the MUG, it becomes obvious that the system itself is capable of more output. The natural question that arises is how can we get more output if needed? However, the MUG not only lets us look at current limitations but also allows a simple means of doing some *what-if* analysis, as shown in question 3a.

Question 3a: If we had the ability to staff a worker at each station in the assembly system with only adding 2 seconds to each workstation for part handling between stations, complete the MUG for question 3a.

See completed MUG for question 3a.

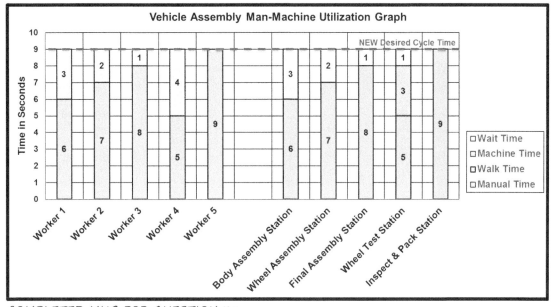

COMPLETED MUG FOR QUESTION 3a

As we can see from the completed MUG for question 3a, the system is capable of producing at a much faster rate if more workers are added. Although adding workers to a system will not always have a dramatic impact, we need to understand the effect that adding labor has to the maximum system output before we can decide if it is feasible to do so. So, again, we ask the question: what is the new maximum system output?

Question 3b: What would be the maximum capacity per hour from the completed MUG for question 3a?

See answer for question 3b.

As shown in the Man-Machine Utilization Chart, worker 5 is the resource with the maximum cycle time for one complete part work cycle. This time is 9 seconds. So, 3600 seconds per hour divided by 9 seconds per one part work cycle = **400 parts per hour.**

ANSWER FOR QUESTION 3b

"SEEING SIMILARITIES IN LIFE."

Question 4: Why is the walk time not shown on the right side of the man–machine utilization chart?

See answer for question 4.

Walk time is not included on the right side of the Man-Machine Utilization Chart because **machines and/or workstations are a part of the physical layout and do not move or walk** in a typical system. Walk time is only tracked with the workers to show the impact of the physical layout on the worker. As changes occur in the system, the walk impact can be captured to see the benefits or cost.

ANSWER FOR QUESTION 4

Question 5: (Choose one) Which of the following items should not be included on a TSS?

1. Pictures and sketches.
2. Key symbols.
3. Step numbers.
4. Major step descriptions.
5. Key points.
6. Reasons.
7. Time.
8. All of the above items should be included on a TSS.

"THERE IS NEVER ONE PERFECT TOOL FOR A SITUATION."

See answer for question 5.

> **The answer is 8.** All of the above items should be included on a Task Summary Sheet. While any of these items may not be captured on a given line, i.e. a key symbol may not be required for a certain step, the fact that Key Symbols should be used where applicable is very important. All of the Task Summary Sheet component items are key to provide a visual, simple and informative standardized work aid for the worker and subsequent work place auditors.

ANSWER FOR QUESTION 5

The TSS is probably one of the most versatile tools since it can be adapted to so many work situations. It is almost tailor-made for applications where the worker must execute a task that is performed on a very infrequent basis. We have used it countless times for ensuring that although a task is performed infrequently, it is executed the same way each and every time, no matter who performs the work on a particular instance. Not only does it help make the task much more predictable and thus reduce variability, it also helps ensure quality—the job is designed to produce the highest quality, and, if we reduce variability, we also reduce the chance of error. Although we cannot guarantee perfection, reducing variation is one way of making the chance of error creeping into our process smaller.

In some standardized work applications, for those using an SWC, for example, most of the steps may be simple and require little explanation. However, this is not always the case as some steps may require standardized work themselves. This is why we call this tool a task summary sheet as it can be used to capture the major steps; important points; reasons why something is to be done a certain way; illustrations; and any other tips, tricks, or information that may be required. It can serve as documentation for steps that require more explanation, which in turn can be used to train new workers as well as preserve the information to make sure that it is not lost—often, this type of information is not captured elsewhere and can be lost over time. Once a worker is trained, he or she normally does not need to access the documentation on these tools unless he or she references them, while he or she is working on an improvement effort. However, when a new worker is brought on board, there must be standardization in the way that he or she is trained in the work, and the TSS along with the other tools can help ensure that this can occur.

As we complete our journey in the field, we would just like to say that we hope that you have found this guide to be useful. We also hope that you enjoyed our attempt at bringing some lightheartedness to this subject. We have had a lot of fun trying to inject some silliness while putting the field guide together. Work does not always have to be so strict and formal.

"GOOD LUCK ON YOUR JOURNEYS IN THE FIELD!"

Reference

Martin, T. D. and J. T. Bell. 2011. *New Horizons in Standardized Work: Techniques for Manufacturing and Business Process Improvement.* Boca Raton, FL: CRC Press/ Taylor & Francis.

Index

This index includes the preface. Page numbers with f refer to figures.

A

Administrative task, 80
Automatic unload, 17, 25, 26–27
Average time, 38; *see also specific stopwatches methods*

B

Best time
   answers to questions, 76–77
   definition, 46, 46f, 47f
   determining, 75
   review questions, 67
   significance, 55

C

Changeovers, 80, 85
Clockwise direction, 12, 14, 106
Communications, 8–9, 35
Continuous improvement
   concept, 29, 124, 142, 151
   DCT, 56
   example, 69–70
   memory stopwatch method #1, 43, 73
   SWC, 34–35
   variation, 76, 117–118
   waste finding, 22
   WCT, 118
Counterclockwise direction, 12, 14, 25, 31–32, 106
Critical customer requirement symbol, 140f
CT_{high}, *see* Highest cycle times (CT_{high})
CT_{low}, *see* Lowest cycle times (CT_{low})
Cues, 39, 62, 71, 116, 124

Current system limit, 134f, 135, 136
Customers
   critical customer requirement symbol, 140f
   requirements, 36, 61, 69–70, 121; *see also* Desired cycle time (DCT); Takt time (TT)
   satisfaction, 35
Cycle times; *see also specific stopwatches methods*
   highest (CT_{high}), 45, 46f
   lowest (CT_{low}), 45, 46f
   machine, *see* Machine cycle times
   observed (OCT)
      answers to questions, 123–124, 126
      OCT < TT, 93, 93f
      OCT > TT, 91, 92f
      review questions, 115–116, 117–118
      significance, 45, 117
      variation and, 46, 47f, 117–118; *see also* Variation (V)
      WCT and, 88
   total, 48–49, 48f, 49f, 64, 134

D

DCT, *see* Desired cycle time (DCT)
Delivery routes, 80
Delphi Corporation, viii
Desired cycle time (DCT)
   answers to questions, 70, 121–122, 123–124
   review questions, 62, 115–116
   significance, 56, 121–122, 123–124
   TT and, 36–37
   total average times and, 53
   WCT, 80, 85, 87f, 89f
Dual nesting, 17, 26–27, 99, 100, 105, 105f

E

Expected benefits, 21

F

Forced wait, 85, 86, 86f, 93, 94f, 135

G

Gemba, ix, 20
General Motors, viii
Geographic sequence and work sequence, 96, 97f
Graphs, 76, 76f; *see also* Man–machine utilization graph (MUG)

H

Hand-off point symbol, 35f, 56f, 140f
Highest cycle times (CT_{high}), 45, 46f
Hospital rule issue symbol, 140f

I

Increased capacity, 21, 22, 29

J

The job instruction, 137

K

Kaizen, vii, 80, 85, 124, 126

L

Lap memories, 43, 48, 55
Layout sketch, 7–17
    basic rules, 9–11
    communication and, 8–9
    creating sketches, 14–16, 23, 29–30
    directionality problem, 11–14
    food for thought, 16–17
    review, 19–32
        answers to questions, 27–32
        review questions, 19–27
    SWC, 35
    unnecessary movement, 7–8
    WCT and, 82–83, 83f, 84f
Leading hand free upon approach, 13–14
Lean, concept, 133
Left handedness, *see* Leading hand free upon approach

Lowest cycle times (CT_{low}), 45, 46f
Lowest repeatable time, 38, 55, 65–66, 75–76

M

Machines
    additional, 17, 26–27, 32
    cycle times, *see* Machine cycle times
    man–machine utilization graph, *see* Man–machine utilization graph (MUG)
    parallel machines #1, 98, 98f
    parallel machines #2, 99–101, 100f, 101f, 102f, 103–105
Machine cycle times
    answers to questions, 122–123, 126–127
    excessive
        parallel machines #1, 98, 98f
        parallel machines #2, 99–101, 100f, 101f, 102f, 103–105
    review questions, 23–25, 115, 118–119
    waiting time and, 24–25
    WCT, 80, 85, 87f, 89f
Man–machine utilization graph (MUG), 134–136
    answers to questions, 152, 153
    current system limit, 135, 136
    function, 135–136
    illustrative example, 134f
    parts, 134–135, 134f
    review questions, 146, 148
    significance, 134, 145–146
    WCT versus, 146
Manual time, 85, 89f, 134f
Maximum capacity, 135, 136, 147, 148, 152, 153
Memory stopwatches
    answers to questions, 72–73
    definition, 37–38
    illustration, 37f
    method #1, 42–45
        answers to questions, 73
        comparison among other methods, 54f
        documentation, 44–45, 45f
        overview, 42–43
        review questions, 63
        steps, 43–44
        summary, 45
    method #2, 48–53
        answers to questions, 74–75
        comparison among other methods, 54f
        documentation, 51–53
        review questions, 65–66
        steps, 49–50

summary, 50
total cycle times, 48–49, 48f, 49f
review questions, 63
summary of all methods, 53–54
Miscellaneous tools, 131–142
introduction, 131–133
man–machine utilization graph (MUG),
134–136; *see also* Man–machine
utilization graph (MUG)
current system limit, 135, 136
function, 135–136
illustrative example, 134f
maximum capacity, 135, 136
parts, 134–135, 134f
significance, 134
review, 145–156
answers to questions, 151–156
review questions, 145–150
supply and demand, 133
task summary sheet (TSS), 137–142; *see also*
Task summary sheet (TSS)
definition, 138
illustrative example, 138f, 139f
the job instruction, 137
parts, 139–141
significance, 138
standardized work, 141–142
standard time, 138
symbols, 140, 140f
time standard, 138
Training Within Industry tool, 137
MUG, *see* Man–machine utilization graph (MUG)

N

New Horizons in Standardized Work (2011),
34, 57, 142

O

Observed cycle times (OCT)
answers to questions, 123–124, 126
OCT < TT, 93, 93f
OCT > TT, 91, 92f
review questions, 115–116, 117–118
significance, 45, 117
variation and, 46, 47f, 117–118; *see also*
Variation (V)
WCT and, 88
Ohno, Taiichi, vii
Output, 21, 22
Overproduction, 93, 93f, 129
Overtime, 29

P

Patient satisfaction issue symbol, 140f

Q

Quality check symbol, 35f, 56f, 140f

R

Read points; *see also* Start/stop points
cues and, 62, 71
memory stopwatch method #1, 43, 44, 45f,
73f
memory stopwatch method #2, 48, 48f, 49,
50, 51, 52f
simple stopwatch method, 40
variation and, 46–47
WCT, 84
Reduced cost, 21
Revenue, 21, 29
Right handedness, *see* Leading hand free upon
approach

S

Safety issue symbol, 35f, 56f, 140f
Security issue symbol, 140f
Simple stopwatches
definition, 37
illustration, 37f
method
answers to questions, 71–72
comparison among other methods, 54f
review questions, 62–63
steps, 40–42
summary of all methods, 53–54
Sound cues, 39, 62, 71, 116
Standard in-process stock summary, 35f, 56f,
140f
Standardized work, vii, 1, 34, 141–142, 149;
see also Man–machine utilization
graph (MUG); Standardized work chart
(SWC); Task summary sheet (TSS);
Work combination table (WCT)
Standardized work chart (SWC), 33–57
characteristics, 35
desired cycle time (DCT), 36–37
food for thought, 55–56
illustrative example, 35f
introduction, 33–34
layout sketch and, 35
purposes, 35

review, 59–77
    answers to questions, 68–77
    review questions, 59–67
standard time, 35, 36, 38
stopwatches, 37–55
    memory stopwatch, 37–38, 37f
    memory stopwatch method #1, 42–45
    memory stopwatch method #2, 48–53
    simple stopwatch, 37, 37f
    simple stopwatch method, 40–42, 63
    start/stop points choosing tips, 38–39
    summary of the three methods, 53–54
symbols used, 56–57, 56f
takt time (TT); *see also* Takt time (TT)
    definition, 35
    DCT and, 36–37
variation (V), 46–47, 46f, 47f; *see also*
        Variation (V)
Standard time; *see also* Desired cycle time
        (DCT); Takt time (TT)
    DCT, 36, 38, 85, 122
    determining, 35, 38, 46, 46f
    significance, 60, 142
    TT and, 122
    TSS, 138
    WCT symbol, 85–86
Start/stop points; *see also* Read points
    choosing, 38–39
    memory stopwatch method #1, 43
    memory stopwatch method #2, 48, 49, 49f
    simple stopwatch method, 40
    WCT, 84–85, 84f
Sterile environment issue symbol, 140f
Stopwatches, 37–54
    memory stopwatch, 37–38, 37f
    memory stopwatch method #1, 42–45
    memory stopwatch method #2, 48–53
    simple stopwatch, 37, 37f
    simple stopwatch method, 40–42
    start/stop points choosing tips, 38–39
    summary of the three methods, 53–54
Straight-lined work cells, 10f
Supply and demand, 133
SWC, *see* Standardized work chart (SWC)
Symbols, 35f, 56–57, 56f, 85–86, 140f

T

Takt time (TT)
    answers to questions, 69–70, 121–122,
        123–124, 126
    average time and, 42
    DCT and, 36–37

definition, 35, 56
formula, 36
review questions, 60–61, 115–116, 117–119
standard time, 122
TT < OCT, 91, 92f
TT > OCT, 93, 93f
variation and, 47f
WCT, 80, 85, 87f, 89f
Task summary sheet (TSS), 137–142
    answers to questions, 154–156
    definition, 138
    illustrative example, 138f, 139f
    the job instruction, 137
    parts, 139–141
    review questions, 150
    significance, 138, 155–156
    standardized work, 141–142
    standard time, 138
    symbols, 140, 140f
    time standard, 138
    Training Within Industry tool, 137
Time standard, 138
Total cycle times, 48–49, 48f, 49f, 64, 134
Toyota, 132
Toyota Production System, 133
Training Within Industry tool, 137
TSS, *see* Task summary sheet (TSS)
TT, *see* Takt time (TT)
Two-handed work, 23–25, 30–31
2-up printed circuit board (PCB), 107

U

Unit of flow, 107–109, 107f, 108f, 109f
Unnecessary walk
    elimination, 22–23
    process steps layout and, 7–8
    review questions, 20, 21
U-shaped work cells, 10f

V

Variation (V), 46–47, 47f
    continuous improvement, 76, 117–118
    memory stopwatch #1, 73
    man and machine, 116–117
    unit flow, 108, 109, 110f
Visual cues, 39, 62, 116, 124

W

Wabash Valley Lean Network, viii
Waiting

forced wait, 85, 86, 86f, 93, 94f, 135
  as waste, 42, 93f, 122–124
Walking
  graphing, 102–103
  MUG and, 154
  significance, 149
  unit flow, 107–108, 110f
  unnecessary
    elimination, 22–23
    layout sketch and, 7–8
    review questions, 20, 21
  walk = zero, 94, 94f
  WCT, 85, 87f, 88f, 89f
Waste; *see also* Waiting; Walking
  reduction, 134, 149
  waiting, 42, 93f, 122–124
  real-life example, 106–109
WCT, *see* Work combination table (WCT)
Wire bonder machine, 99–101, 100f, 101f, 102f,
    103–105
Work cells, 10–11, 10f, 26–27, 32
Work combination table (WCT), 79–110
  food for thought, 105–106
  introduction, 79–82
  man–machine utilization graph (MUG)
      versus, 146
  methodology, 82–91
    step 1: layout sketch, 82–83, 83f
    step 2: draw rough diagram, 83, 84f
    step 3: define read points and capture
        times, 84–85, 84f
    step 4: enter data into WCT form, 85, 86f

    step 5: create work combination
        diagram, 85–87, 86f, 87f, 89f
    step 6: use WCT to look for
        opportunities, 87–88, 89f, 90f, 90
  normal, 91, 91f
  real-world problems, 91–102
    forced wait, 93, 94f
    OCT < TT, 93, 93f
    OCT > TT, 91, 92f
    parallel machines #1 (multiple passes),
        98, 98f
    parallel machines #2 (outsmarting
        ourselves), 103–105
    parallel machines #2 (wire bonder
        machine), 99–101, 100f, 101f, 102f
    walk = zero, 94, 94f
    worker returns to same location,
        96, 97f
    work sequence and geographic
        sequence, 96, 97f
  review, 113–129
    answers to questions, 120–129
    review questions, 113–119
  rules and guidelines, 102–103, 102f
  significance, 80
  summary, 109, 110
  waste, 106–109
Work elements, 84–85, 84f, 94, 94f, 96f
Workers, 96, 97f, 124
Work sequence and geographic sequence,
    96, 97f
Workstations, *see* Work cells

About the Authors

TIM

Timothy D. Martin is a process engineer with more than 30 years of experience in manufacturing engineering and, for the last six years, has been working in the healthcare industry. He has been a Lean practitioner for more than 20 years and brings with him a broad range of continuous improvement experience including extensive hands-on Lean implementation in the United States as well as coaching and mentoring Lean implementations in several countries during his career. In addition to his Lean experiences, he has led hundreds of successful continuous improvement projects in North America as well as developed and delivered Lean training at sites in the United States, Mexico, Europe, and Asia. Tim is currently one of the leaders in the Lean transformation of a healthcare system with 14 hospitals and multiple physician practices.

Tim earned a bachelor of science in electrical engineering technology degree from Purdue University, a master of science in management degree from Indiana Wesleyan University, and an associate in computer science degree from Brewer State Junior College. He received training in the Toyota Production System from the Toyota Supplier Support Center where he was also validated as a trainer in standardized work.

During his many years in the automotive electronics industry, Tim was a staff-level engineer, an engineering supervisor, a Lean manufacturing systems leader, and a strategic planner. An innovative and creative engineer, he received multiple patents and defensive publications. Tim has provided technical expertise for several mergers, acquisitions, and joint ventures. He is a certified Black Belt in Lean Six Sigma healthcare through Purdue Healthcare Advisors and holds a Lean Bronze Certification through the Society of Manufacturing Engineers. In his early career, he spent several years in the electrical manufacturing industry.

JEFF

Jeffrey T. Bell has more than 30 years of experience in manufacturing. He started in 1982 working in the aviation/aeronautical powerplant and electromotive component manufacturing and assembly arena. In 1991, he shifted into the automotive sector and worked in various areas of circuit board and integrated circuit fabrication and assembly for automotive electronics in the United States, Mexico, and Europe. Jeff was introduced to Lean in the 1990s and has been a practitioner ever since. In 2012, Jeff moved into the commercial kitchen and bath cabinetry industry where he spent two years as the plant's Lean manager resurrecting their Lean culture. He is currently working at Franciscan St. Elizabeth Health in the business transformation group as one of the leaders in the Lean transformation of the Franciscan Alliance and their 14 hospitals and multiple physician practices.

Jeff has spent his whole career of project work, operations, and product line support involved in the fields of industrial engineering and Lean manufacturing. His journey started as a floor support industrial engineer and has had various industrial engineering (IE) assignments in core IE, preproduction/product design, supplier development, management, leadership, and new plant start-up. The countless number of hours of time spent observing in the *gemba* have given him much of his perspective in understanding standardized work and its supportive impact on safety, quality, delivery, and cost with respect to process, people, and business. His work has allowed him to meet several hundred great people during the combined efforts of continuous improvement throughout his career.

Jeff earned a bachelor of science in industrial engineering degree from Kettering University and a master of science in industrial engineering degree from Purdue University.

Tim Martin and Jeff Bell are also coauthors of *New Horizons in Standardized Work: Techniques for Manufacturing and Business Process Improvement* (2011).

SCOTT

Scott A. Martin is a freelance artist/illustrator and business manager living in Birmingham, Alabama. His educational background includes studying general liberal arts at Beville State Community College in Fayette, Alabama as well as studying communications at the University of Alabama in Tuscaloosa, Alabama; however, he is artistically self taught. His range of influences includes Frank Frazetta and Bill Watterson. His unique styling is a culmination of over 35 years as an artist. His artistic mission is to remain true to his own styling while drawing a hard line between inspiration and imitation. Although his most current works are in digital media, he is also experienced with other forms of traditional art including pencil, ink, oil, and acrylics. He is also an amateur sand sculptor, pumpkin carver, and printmaker. His most recent works include book illustrations and website designs as well as private art collections throughout the United States.